AF849997

Maria Rikitianskaia

The Global Wireless

History and Ideas

New Perspectives in European Studies

Edited by
Fernanda Gallo, Florian Greiner and Jan Vermeiren

Volume 3

Maria Rikitianskaia

The Global Wireless

Transnational Radiotelegraphy and Its Disruption in World War I

DE GRUYTER
OLDENBOURG

The open access publication of this book has been published with the support of the Swiss National Science Foundation.

ISBN 978-3-11-120130-6
e-ISBN (PDF) 978-3-11-120232-7
e-ISBN (EPUB) 978-3-11-120301-0
ISSN 2750-1493
DOI https://doi.org/10.1515/9783111202327

This work is licensed under the Creative Commons Attribution-NonCommercial-NoDerivatives 4.0 International License. For details go to https://creativecommons.org/licenses/by-nc-nd/4.0/.

Creative Commons license terms for re-use do not apply to any content (such as graphs, figures, photos, excerpts, etc.) not original to the Open Access publication and further permission may be required from the rights holder. The obligation to research and clear permission lies solely with the party re-using the material.

Library of Congress Control Number: 2024939730

Bibliographic information published by the Deutsche Nationalbibliothek
The Deutsche Nationalbibliothek lists this publication in the Deutsche Nationalbibliografie; detailed bibliographic data are available on the internet at http://dnb.dnb.de.

© 2024 the author(s), published by Walter de Gruyter GmbH, Berlin/Boston
The book is published open access at www.degruyter.com.

Cover image: ITU map of radiotelegraph channels, 1927, © International Telecommunication Union
Typesetting: Integra Software Services Pvt. Ltd.
Printing and binding: CPI books GmbH, Leck

www.degruyter.com

Acknowledgements

This book took many years in the making. It started as a PhD dissertation, successfully defended in 2018 at Università della Svizzera italiana, Lugano, Switzerland. In 2019, the dissertation was awarded the 2019 Turriano ICOHTEC Prize, the Early Career Prize for Books on the history of technology, Then, the manuscript was then revised and rethought from a more contemporary angle because of another project about contemporary Wi-Fi that I led at London School of Economics and Political Sciences, London, UK. After that, as a Byrne-Bussey Marconi Fellow in the History of Science, Technology and Communication at the Bodleian Library in 2022-2023, Oxford, UK, I revised this research even more in light of other different archival sources. It is finally published when I work at Regent's University London. During the research and my work in these various places, I met accomplished media scholars, researchers, historians, radio amateurs, writers and journalists from whom I received valuable advice that influenced my ideas and the ways of thinking about new technologies. I am grateful to all of them for shaping my ways of thinking about early communication technologies.

Thank you very much for the support and help, endless discussions and valuable advice, insights and input, along with doubts and comments, throughout the years of working on this book: Gabriele Balbi, Dylan Mulvin, Simone Natale, Hugh Chignell, Dwayne Winseck, Kate Lacey, Alec Badenoch, Fernanda Gallo, Andreas Fickers, David Hendy, Marc Raboy, Jamie Medhurst, Sabina Mihelj, Alena Pfoser, Katharina Lobinger, Matthew Hibberd, Paolo Bory, Gianluigi Negro, Eleonora Benecchi, Gloria Dagnino, Elena Gargaglia, Zhan Zhang, Marco Cucco, Philip di Salvo, Ester Lo Biundo, Lily Zemnukhova, Nikolay Rudenko, Natalie Rangelov, Marlen Heide, Lakshmi B. Nair, Rukhshona Nazhmidinova, Kübra Canhilal, Marie Cronqvist, Dick van Lente, Simone Fari, Elizabeth Bruton, Graeme Gooday, Richard R. John, Juraj Kittler, Simon Potter, Valérie Schafer, Christian Schwarzenegger, Manuel Menke, Martin Thom, Anna Åberg, Tirza Meyer, Ana Paula da Silva, George Nikolaenko, Michail Maiatsky, Irina Gluschenko, Gayane Kuryatova, Tatiana Bakina, Alina Volynskaya, Anastasia Filcheva, Ekaterina Genovese, Julia Weiner, Catherine Davidson, Mireille Hebing, Ernesto Gallo, Deborah Schultz, Irina Miles, Stephen Barber, Angels Trias-I-Valls, Victoria Pastor-Gonzalez, Vanessa King and many others.

I am extremely grateful for the help of the archivists. The ITU Library and Archives provided their generous support in my research and work, particularly Heather Heywood, Kristine Clara and Susanne Schaefer. The curator of Dokumentationsarchiv Funk in Vienna, Wolf Harranth, not only showed me the documents, books and QSL cards but also opened a world of radio amateurs' culture. I thank Michael Hughes, Rachel Naismith and Alexandra Franklin for their guidance

Open Access. © 2024 the author(s), published by De Gruyter. (cc) BY-NC-ND This work is licensed under the Creative Commons Attribution-NonCommercial-NoDerivatives 4.0 International License.
https://doi.org/10.1515/9783111202327-202

through the extensive collection of the Marconi archive in Bodleian Library, Oxford, during my Byrne-Bussey Marconi Fellowship. I am also grateful to Jon Cable from the archives of the Institution of Engineering and Technology and James Elder from the BT Archives in London. In addition, I would like to thank those archivists and librarians who supported me throughout my research in finding bits and pieces of the biggest picture from unexpected sources, especially the staff of the British Library and the Russian State Library.

I am grateful to the Swiss National Science Foundation (SNSF) for supporting my research and mobility projects over the years. I thank deeply Università della Svizzera italiana, London School of Economics, Loughborough University, Bodleian Library and Regent's University London for hosting me and this project. I also thank scholars of the Tensions of Europe network, the Italian Society of Science and Technology Studies (STS Italia), Entangled Media Histories Network (EMHIST), the International Committee for the History of Technology (ICOHTEC), the Media History division of the International Communication Association (ICA), European Communication Research and Education Association (ECREA), and many other networks.

Last but not least, I am extremely grateful to my family for their unconditional support and for having my back: my parents, Prof. Yuriy Vaguine and Lidia Vagyna; Nelli Rikitianskaia, Liubov Rikitianskaia and Mikhail Rikitianskii; and, of course, my husband Andrei Rikitiankii and our children, Alexander Rikitianskii and Ekaterina Rikitianskaia, without whom this book would have never been accomplished.

Thank you.

Contents

Introduction

The world has been experiencing a wireless boom, with every technological innovation of recent years emphasizing its novel wireless status. Technology companies announce launches of wireless earbuds, wireless chargers, and wireless video transmission systems. Domestic appliances are not left behind, with advertisements for wireless hoovers, wireless home security, and even wireless window blinds. The beauty industry parades with wireless oral irrigators, wireless curling wands, and wireless cleansing brushes. The beginning of the twenty-first century seems to be obsessed with wireless.

Essentially, there are two ways of turning a technology into a wireless one. The first approach is to replace a cord with radio antennas, so that the two elements are still able to communicate and connect instantly. For example, wireless headphones will receive a signal from a smartphone to play music instantly. Another option is to substitute simple batteries for electric wires, as in the case of oral irrigators or hoovers. No communication in the moment is taking place, yet everything with an electric portable charge could in fact be described as 'wireless.'

This book explores the first, communicative type of wireless, which takes the form of radio technology. This type of wireless connectivity was the most common up until recent times, and in fact the Oxford English Dictionary defines wireless as 'a device, system, or communications network which uses radio waves, microwaves, etc., for the transmission and reception of signals, rather than a physical connection.'[1] Devices sending and receiving radio signals surround us on all sides. They lurk in our bags or wallets, accompany us every step of the way as we move through our urban environments, and are an indispensable part of our everyday activities. They are smartphones, computers, earphones, but also contactless credit cards, garage door openers, TV remote controls, and many other such things. Wireless has conquered the world.

How did this wireless world come into being? Wireless history starts with the discovery of radio technology in the late nineteenth century. It was a groundbreaking invention for society. Short and long electrical pulses allowed users to code any message into dots and dashes and send it over electromagnetic waves. This meant that without any cables, on a relatively portable device, any person could transmit and receive messages as part of a large media network. A range of different experiments with the use of radio waves took place over several decades and in the 1920s they reached their peak in the spread of the well-known medium: radio broadcasting as we know it today.

1 "Wireless, Adj. and n., Oxford English Dictionary" (Oxford University Press, 2020), https://www-oed-com.proxy.sbu.usi.ch/view/Entry/229458.

Open Access. © 2024 the author(s), published by De Gruyter. BY-NC-ND This work is licensed under the Creative Commons Attribution-NonCommercial-NoDerivatives 4.0 International License.
https://doi.org/10.1515/9783111202327-001

The experiments with radio did not stop with the spread of radio broadcasting. The twentieth century witnessed the invention of yet other radio technologies: radiography, radiotelephony, radar, radiomobile, packet radio networks, and others. As far as twentieth-century inventions were concerned, the word 'radio' was as influential a keyword as 'wireless' is today.

These wireless innovations were indispensable for the progress of media and communications. Wireless and radio technologies today form a valuable part of communications infrastructure, allowing us to easily connect when on the move, enhancing the portability and general user-friendliness of the devices. Without radio technologies, the cities of today would be overwhelmed by a forest of cables and wires. As early as the late nineteenth century, there was already a big problem with cables. The telephone tower in Stockholm at that time connected 5,000 telephone lines at once and looked as if it were entangled in a spiders' web (see Figure 1). Growing communication connectivity could well have struggled for decades with the vulnerability and complexity of developing cable infrastructure, had it not been for wireless.

Figure 1: A Stockholm tower connecting 5,000 telephone lines at the end of the nineteenth century.[2]

2 Tekniska Museet, "Telefontornet i Rimfrost, 1890-Talet. Stockholms Allmänna Telefon" (1890), https://digitaltmuseum.se/021016458638/telefontornet-i-rimfrost-1890-talet-stockholms-allmanna-telefon-ab.

The invention of radio technology in the late nineteenth century was truly a godsend as far as the problem of wired and cable infrastructure was concerned. However, wires still remain a very important part of this communications infrastructure, and cables do today transport most of the Internet traffic. The highly physical infrastructure operative today, such as servers, fibre optics, and warehouses constitutes the most important part of the IT market.[3] Yet user experience is defined by the last link in the chain and, today, for many users this last link is wireless. Connected to their Wi-Fi routers at home, the final users rely on the connection in the air. We are so used to being mobile and connected, and not being attached to any cable, that a very important part of the electronic market today is in fact concerned with Wi-Fi extenders – the radio signal repeaters that allow the network to extend to the remotest corners of a house. Wireless is an indispensable part not only of the global communications infrastructure, but also of the domestic environment and routine everyday experience.

This book explores the early history of wireless networks. *The Global Wireless* charts a history of wireless beginning in the 1910s, when it was used as a tool for global communication, and ending as it declined and slowly fell out of view in the 1920s. Contributing to media history and science and technology studies, *The Global Wireless* recounts how the advent of wireless technologies created a novel socio-technical problem; since radio signals easily and unwittingly crossed national borders, they challenged existing systems and standards of national media infrastructure control. The book further examines the political negotiations around the International Telecommunication Union, the growth of international communication networks, and the expansion of global media companies on the eve of World War I. *The Global Wireless* demonstrates that long before Wi-Fi and 5G, another wireless technology had already spread around the globe and prompted, in its wake, a radical reconsideration of networked communication and community.

Based on previously unexplored archival materials, *The Global Wireless* reverses the usual approach to the origins of radio broadcasting by showing that in the early twentieth century wireless technologies were used for global connectivity rather than being confined to the status of national media. The book demonstrates that in the early twentieth century the global networking potential of wireless was ultimately disrupted by the exigencies of national security during World War I, and hence forged a long-enduring understanding of radio as a na-

3 Nicole Starosielski, *The Undersea Network* (Durham: Duke University Press, 2015); Nicole Starosielski and Lisa Parks, eds., *Signal Traffic: Critical Studies of Media Infrastructures* (University of Illinois Press, 2015).

tional mass medium. Scrutinizing international technical standards organizations, complex infrastructures, and a transnational constellation of individual actors, *The Global Wireless* demonstrates how global wireless networks – despite being destroyed by war – nonetheless produced social and technical reverberations felt throughout the twentieth century and into the wireless world of today.

The Global Wireless tells a transnational history of wireless. Radio waves could hardly be restricted by national boundaries, and therefore wireless telegraphy (also called radiotelegraphy) was inherently a transnational medium from the very beginning. Accordingly, the development of wireless telegraphy necessitated both transnational regulations and negotiations, which also became a matter of lengthy discussions in international arenas, the International Telecommunication Union among them. The book traces the transformation of wireless over the course of World War I, the event that accelerated the transnational exchange of knowledge, people, goods, information on a global scale. This monograph therefore recounts a transnational history of a medium by scrutinizing a transnational object (radio), in a transnational setting (war), with a particular focus on transnational negotiations and regulations in a specific transnational institution (the ITU).

The aim of the book is twofold. First of all, it seeks to present a complex account of intertwined national and international influences and relationships on the European wireless scene. The monograph thus provides an overview of several communication projects designed for global use, such as international time signals and international meteorological reports, and reveals how World War I disrupted the communications infrastructure and stimulated the development of nationalized radio. This perspective enables the book to bridge the seemingly distant histories of telecommunications and mass media, global and national communication, *la belle époque* and the roaring twenties.

Second, this book provides historical background to current wireless policy. Wireless telegraphy in the 1910s allowed anyone—from a highly qualified professional to an outright amateur—to be 'on air' and both to transmit and receive messages as part of a broader global community, thus anticipating and even outpacing the participatory culture created in the digital environment of the 1990s and 2000s. It also predated many globalization phenomena of communication. The problems discussed and solved at the beginning of the twentieth century, such as those of private and public management, the radio spectrum and the involvement of amateurs, paved the way for many later political decisions and even today still reflect the perennial struggles of the wireless industry. The security of the global networks was highly problematic at the start, as indeed it has continued to be up until the present day. Understanding how innovations were handled, understood, regulated, imagined, and framed in the past can help to shed light upon the current digital revolution.

The research places Europe at its center: with competitive European inventors and engineers, radio waves easily crossing borders, and great damage and suffering inflicted by the war. It investigates the transformation of wireless in Europe that took place over the course of World War I. In particular, this monograph studies the transformation of wireless, as developed in the European empires, from the 1912 International Conference on Wireless Telegraphy in London (which laid down the ground rules for the international use of the medium) to the Washington conference of 1927 (which dealt with the consequences of the war and regulated radio broadcasting). The book addresses the following research questions: How and why did European radio develop a broadcasting function and fall into the role of a one-to-many medium? How and why did the transnational dimension of radio come to be transformed between 1912 and 1927? How and why did World War I affect the evolution of radio? What influence did the international arena, corporate strategies, technical expertise, and amateur experimentation have on the development of radio?

The main goal of this book is to demonstrate that a drastic change in the use of radio technology did indeed occur over the course of World War I, but that it was not merely a turn to radio broadcasting as earlier scholarship supposed. Even though the activities of broadcasting stations in the 1910s were irregular and sporadic, radio broadcasting was an already established practice and was indeed born before the war as a transnational medium. This research concludes that the war served to disrupt the transnational networks of information exchange and brought to the fore a novel idea of radio as a predominantly national medium uniting the people of a particular country. This idea totally replaced the pre–war vision of radiotelegraphy as a transnational medium shared by different actors, such as international arenas, corporations, experts, and amateurs. Overall, the monograph demonstrates that national radio was a natural enough development of the technology, but only arose as a consequence of the World War I.

The transnational history of wireless

Doing transnational research

The novelty of this book lies in the decision taken to address radio and wireless history from a transnational perspective. The transnational approach helps to identify novel phenomena and dependencies that cannot be grasped within a national framework, such as knowledge and technology transfer and exchange across national borders. Iriye and Saunier state that a transnational perspective invites us to consider links and flows of 'people, products, processes and patterns that operate over, across, through, beyond, above, under, or in- between polities and societies.'[4] Over the past decades, transnational history has flourished, with some scholars even going so far as to speak of a 'transnational turn' in historiography.[5] Scholars underscore the importance of broadening 'one's presumed object of study' and going beyond traditional boundaries in radio history.[6] Media histories have typically been written from national perspectives, reflecting the manner in which government records are kept. However, recent studies have shown the importance of adopting a global and transnational approach, one that would allow us to look more broadly at the origins of radio broadcasting.

The use of the term 'transnational' was sparse prior to its spread in the political science of the 1970s.[7] Etymological scrutiny of the term indicates that in a linguistic context it referred to research that focused on the families of those who were bilingual and, in the US, it was also used as a synonym for 'transcontinental' or traversing the whole nation.[8] A transnational approach was also employed in the course of re-examining national or local histories, when encountering multiple cultural meanings relating to the construction of a national identity; consider, for example, the research on the Gotthard Railway tunnel and the inscription of this engineering project into Swiss national identity.[9] Hans-Ulrich Wehler has noted that the meanings of a transnational approach in different fields of historical scholarship are distinct and depend upon the angle of the research.[10] There are at least three important concepts that these different transnational histories share: nation-state, flows, and borders.

---

4 Akira Iriye and Pierre-Yves Saunier, "Introduction: The Professor and the Madman," in *The Palgrave Dictionary of Transnational History. From the Mid-19th Century to the Present Day*, ed. Akira Iriye and Pierre-Yves Saunier (New York: Springer, 2009), xviii.

5 Erik Van der Vleuten, "Toward a Transnational History of Technology: Meanings, Promises, Pitfalls," *Technology and Culture* 49, no. 4 (2008): 974.

6 Noah Arceneaux, "The Wireless Press and the Great War: An Intersection of Print and Electronic Media, 1914–1921," *Journal of Radio & Audio Media* 26, no. 2 (2019): 330, https://doi.org/10.1080/19376529.2018.1497035.

7 Simon MacDonald, "Transnational History: A Review of Past and Present Scholarship" (UCL Centre for Transnational History, 2013).

8 Van der Vleuten, "Toward a Transnational History of Technology: Meanings, Promises, Pitfalls."

9 Judith Schueler, *Materialising Identity: The Co-Construction of the Gotthard Railway and Swiss National Identity* (Amsterdam: Aksant, 2008).

10 Hans-Ulrich Wehler, "Transnationale Geschichte — Der Neue Königsweg Historischer Forschung?", in *Notizen Zur Deutschen Geschichte* (München: Verlag C.H. Beck, 2007), 61–73.

First of all, transnational history challenges the idea of the nation-state as the principal organising category in academic research. Initially, the term 'transnational' addressed contacts, coalitions, and interactions across state boundaries where at least one of the participants was a non-national actor.[11] Transnational history is considered to be an approach that was developed in order to research multi-influences that lie beyond the terms of a nation- or state-centered history. This method has even been criticized for running the risk of neglecting the power of national parties, given its re-centring of the focus from a national framework to a transnational one.[12]

Another constitutive element of transnational history is its concern with flows.[13] The question of flows has been discussed extensively in media and communication studies as well as in the social sciences and humanities. In *The Information Age*, Manuel Castells[14] wrote that 'our society is constructed around flows: flows of capital, flows of information, flows of technology, flows of organizational interactions, flows of images, sounds, and symbols.' He defined flows as 'purposeful, repetitive, programmable sequences of exchange and integration between disjointed positions held by social actors in the economic, political, and symbolic structure of society.'[15] The main idea behind this preoccupation with flows is to bring into focus 'open-ended multiple interconnectivities of distant locales and identities.'[16] Research into flows has revealed emerging and transversal dependencies among different groups, actors, and spaces. According to Thussu, flows differ by their very nature and may be divided into three broad categories: global, transnational, and geo-cultural.[17] This typology is most fully applicable to the contemporary mediascape. The global, or 'dominant,' flows are mainly concentrated around big media corporations that have an impact on a global audience, such as Disney, Google, and the BBC. Transnational and geo-cultural flows represent a

---

11 Robert O. Keohane and Joseph S. Jr. Nye, "Transnational Relations and World Politics: An Introduction," *International Organization* 25, no. 3 (1971): 329–49.

12 Van der Vleuten, "Toward a Transnational History of Technology: Meanings, Promises, Pitfalls."

13 Simon MacDonald, "Transnational History: A Review of Past and Present Scholarship" (UCL Centre for Transnational History, 2013).

14 *The Information Age: Economy, Society, and Culture. Volume I. The Rise of the Network Society* (Oxford: Blackwell, 1996), 442.

15 Ibid., 442.

16 Miyase Christensen, "Trans National Media Flows: Some Key Questions and Debates," *International Journal of Communication* 7 (2013): 2400–18.

17 Daya Kishan Thussu, "Mapping Global Media Flow and Contra-Flow," in *Media on the Move: Global Flow and Contra-Flow*, ed. Daya Kishan Thussu (London and New York: Routledge, 2007).

group of 'contra-flows,' which exert regional or local influence.[18] In historical research, scholars usually refer to transnational flows as flows over national borders, even though they may also affect global or regional audiences.

The next critical issue regarding flows is the topic of borders and boundaries. Definitions of transnational history embrace the idea of movements, trends, and forces that deal with 'transcending the national' in a physical sense and cutting across boundaries.[19] These borders are not necessarily identical with state borders because a nation can be defined in terms of many different characteristics. In this sense, a transnational approach also engages in dialogue with border studies and human geography. The concern to grasp the constructed nature of borders, and to view borderlands as flexible spaces, has led in recent decades to the carving out of a flourishing new field of study. According to Jones, 'instead of demonstrating that boundaries and categories are fixed, and natural, the renewed interest in boundaries exposes their inchoateness.'[20] Transnational history addresses borders from a connection perspective. In this sense, the concept of borders stands now in opposition to that of 'frontiers,' the former reflecting expanding socioeconomic world-systems rather than national peculiarity.

The transnational approach stemmed from critiques of comparative history, and one such by Marc Bloch is often cited as a key source.[21] Comparative history involves the comparison of different societies that share similar cultural conditions or co-exist in the same period. It is often considered a nation-centered approach, even if it may also account for several different national frameworks. This type of history has been criticized for its inability to transcend the boundaries of national historiography because even in a comparison, the national units in question will be bound up with the national framework. Comparative history has also been criticized for its neglect of variations in space and time of national experiences, thus rendering it difficult or even impossible to research cases to do with borders. In particular, any continuity or mutual influence between the different nations would have to be neglected when setting up appropriate cases for comparison, as 'the comparison breaks continuities, cuts entanglements, and interrupts the flow of narration.'[22] The Annales School, a French twentieth-century

18 Ibid.

19 Van der Vleuten, "Toward a Transnational History of Technology: Meanings, Promises, Pitfalls," 988.

20 Reece Jones, "Categories, Borders and Boundaries," *Progress in Human Geography* 33, no. 2 (2009): 89, https://doi.org/10.1177/0309132508089828.

21 Simon MacDonald, "Transnational History: A Review of Past and Present Scholarship" (London, 2015).

22 Jürgen Kocka, "Comparison and Beyond," *History and Theory* 42, no. 1 (2003): 39–44.

school of historians, demonstrated that where long-term social history was concerned, national units could not suffice as frames of analysis, an idea which necessitated a new understanding of the transnational. In contrast to comparative history, the transnational approach does not presuppose that objects can be easily distinguished but rather that they mutually influence each other.

There are a number of other, related but still distinct approaches, alongside the transnational approach, which pay attention to flows over national borders and question the national framework. One such approach is transfer studies, which was elaborated upon by Michel Espagne and also investigates the history of knowledge transfer across different cultural contexts.[23] Transfer studies thus emphasize how national cultures are created through the transfer of knowledge across national and cultural boundaries, however, the problem here is that the transfer is still considered within a national framework: from one entity to another.[24] The French tradition of *histoire croisée* claims to have solved this problem. According to this perspective, national paradigms undergo constant economic, technological, and cultural transfer and are always in the process of formation thanks to their intersections (or intercrossings, in other translations). However, apologists of *histoire croisée* Werner and Zimmermann claim that this multidimensional approach considers objects not in relation to one another but instead through one another.[25] Entangled history is another, related approach, one that derives from Mintz's study on the history of sugar[26] which focuses on the links and exchanges between different regions of the world. Entangled history likewise criticizes comparative histories but in contrast to the abovementioned historiographical traditions it examines entanglements between quite distant units. It also underscores the fact that transfer processes run not only from the position of power downwards to weaker, subaltern actors but also in the opposite direction as well. These approaches, in contrast to a transnational approach, are often criticized for their overly fixed frames of reference and for their propensity to ignore processes of transformation at the border, the direction of the transfer and how precisely overall, constant change connects to the transfer. As a consequence, the units of analysis deployed by those following these approaches tend to become frozen or invariable.

23 Michel Espagne, "La Notion de Transfert Culturel," *Revue Sciences/Lettres* 1, no. 1 (2013), https://doi.org/10.4000/rsl.219.

24 Michael Werner and Bénédicte Zimmermann, "Penser l'histoire Croisée: Entre Empirie et Réflexivité," *Annales. Histoire, Sciences Sociales* 58, no. 1 (2003): 7–36.

25 Michael Werner and Bénédicte Zimmermann, "Beyond Comparison: Histoire Croisée and the Challenge of Reflexivity," *History and Theory* 45, no. 1 (2006): 30–50.

26 Sidney Mintz, *Sweetness and Power: The Place of Sugar in Modern History* (Harmondsworth: Penguin Books, 1986).

Global history is yet another approach related to transnational history. A concept that emerged in the 1990s after the debates then raging on 'globalization,'[27] it focuses on supra-national phenomena, such as economic globalization, environmental issues, and international relations, examining connections rather than flows and the fluid character of events, phenomena, and products.[28] Moreover, historically, global history has commonly been used to describe the recent past, while a transnational approach frequently studies the nations' pasts through their interactions, flows, and collaborations.[29] The crucial distinction between a transnational approach and global history is that the latter tends to investigate global phenomena instead of acknowledging the flexibility and fluidity of national frameworks.

All these approaches, despite their different labels, remind us that the classical version of comparative history has proved too formalized and rigid, and has suffered from a tendency to neglect the relational and fluid character of the units of comparison employed.[30] According to some scholars, the differences between these approaches are not essential, unless a researcher is specifically concerned with the constraints and benefits of an enlargement of the focus. For instance, Espagne, who himself coined the term 'cultural transfer' and is believed to be one of the founding fathers of transfer studies, made a significant contribution to the transnational approach, and recently claimed that transfer studies are part and parcel of transnational cultural historiography.[31] All these approaches respond to a larger trend in scholarship, involving a concern 'to reconstruct aspects of the human past that transcend any one nation-state, empire, or other politically defined territory.'[32]

Transnational communication research

With the invention of electricity, information began to travel easily beyond national frontiers, with this cross-border exchange drastically transforming global markets. The global media system developed in the period between the nine-

27 C.A. Bayly et al., "AHR Conversation: On Transnational History," *The American Historical Review* 111, no. 5 (December 2006): 1441–64.
28 Caroline Douki and Philippe Minard, "Histoire Globale, Histoires Connectées: Un Changement d'échelle Historiographique," *Revue d'histoire Moderne et Contemporaine* 5, no. 54 (2007): 7–21.
29 Akira Iriye, "The Rise of Global and Transnational History," *Journal of Transnational American Studies* 5, no. 1 (2013).
30 Andreas Fickers and Catherine Johnson, "Transnational Television History: A Comparative Approach," *Media History* 16, no. 1 (2010): 1–11, https://doi.org/10.1080/13688800903395411.
31 Espagne, "La Notion de Transfert Culturel."
32 Bayly et al., "AHR Conversation: On Transnational History."

teenth century and the 1920s.[33] The global electric telegraph was historically one of the most researched in regard to its global connectivity, along with other cable-based telecommunication networks that require massive financial investment. However, as Topik and Wells have asserted, along with the earlier telegraph and telephone, the radio was also 'a communication system that demanded large investments in coordinated networks.'[34] This transnational knowledge and information exchange was the key to the development of radiotelegraphy. It was a transnational transfer of knowledge that allowed physicists and experimenters from around the world to develop radio, a 'detour from Karlsruhe via Geneva, Paris, Liverpool, Krontstadt and Bologna to London.'[35]

Since radio waves deal with information flows, new phenomena, and causal relations that transcend national spaces, a transnational approach enriches the history of radio broadcasting and wireless telegraphy because it focuses on the constant exchange of knowledge, techniques, and news across national borders.[36] As Fickers explains, historians were tempted to analyze mass media from a national perspective due to their national institutional structure; archives had been assembled by national organizations, preserved in their national languages, and supported the idea of a nation. National research on media often reinforced the idea of Benedict Anderson about the 'imagined communities' created by media.[37]

Following previous research on transnational histories of broadcasting since 1925,[38] this monograph proposes a transnational perspective on radiotelegraphy

---

33 Robert M. Pike and Dwayne R. Winseck, "The Politics of Global Media Reform, 1907–23," *Media, Culture & Society* 26, no. 5 (2002): 643–75, https://doi.org/10.1177/0163443704045505; Dwayne R. Winseck and Robert M. Pike, *Communication and Empire: Media, Markets, and Globalization, 1860–1930* (Durham: Duke University Press, 2007).

34 Steven C. Topik and Allen Wells, *Global Markets Transformed, 1870–1945* (Boston, MA: Harvard University Press, 2012), 91.

35 Wolfgang Hagen, *Das Radio: Zur Geschichte Und Theorie Des Hörfunks – Deutschland/USA* (Munich: Fink, 2005), 46.

36 Alexander Badenoch and Andreas Fickers, "Introduction: Europe Materializing? Toward a Transnational History of European Infrastructures," in *Materializing Europe: Transnational Infrastructures and the Project of Europe* (Basingstoke, Hampshire and New York: Palgrave Macmillan, 2010), 1–26.

37 Andreas Fickers, "Seeing the Familiar Strange: Some Reflections about Actants, Actors and Arenas of Transnational Media History," *Medien & Zeit. Kommunikation in Vergangenheit Und Gegenwart* 4 (2011): 16–24.

38 Andreas Fickers and Suzanne Lommers, "Eventing Europe: Broadcasting and the Mediated Performances of Europe," in *Materializing Europe: Transnational Infrastructures and the Project of Europe*, ed. Alexander Badenoch and Andreas Fickers (Basingstoke, Hampshire and New York: Palgrave Macmillan, 2010); Suzanne Lommers, *Europe – on Air: Interwar Projects for Radio Broadcasting* (Amsterdam: Amsterdam University Press, 2012); Simon J. Potter, *Wireless Internationalism*

during the 1910s and 1920s. In particular, this research combines three ways of creating a transnational history as described by Van der Vleuten: it researches cross-border flows, looks at international organizations shaping the modern world, and looks at the nation-state from its purportedly central role as the key organizing category.[39] More precisely, this research investigates (1) the history of radiotelegraph as a transnational media, meaning research into transnational cross-border information, technology and actor flows, (2) the international media institutions that impose regulations and shape media development, and (3) global media corporations or actors that drive economic and societal change by implementing innovations, particularly companies and radio amateurs.

The radiotelegraph was a type of transnational media. Brüggermann and Schulz-Forberg have developed a typology for four different types of transnational media: (1) national media with a transnational mission, (2) international media that function in the guise of cooperation between two or more nations and address two or more national audiences, (3) pan-regional media that address a specific world region, and (4) global media that are not restricted to a specific world region but instead address a broad transnational audience.[40] Radiotelegraphy falls into the fourth ideal type of transnational media; it facilitated communication between nations, and the international projects such as time signals, meteorology, wireless navigation were global media projects developed from the cooperation of several nations on the international political scene, through institutions such as the ITU, to address a broad transnational audience. This is a quite drastic difference with radio broadcasting studies, because transnational research into the history of radio typically involves either the first or the third category deployed by Brüggermann and Schulz-Forberg. Research into national media with a transnational mission (the first category) or modifications of this ideal type are quite common among historians of radio broadcasting, as with the history of the BBC foreign-language service that aims to reach an audience outside the national territory, with previous studies highlighting this kind of transnational media history.[41] In addition, regard-

and Distant Listening: Britain, Propaganda, and the Invention of Global Radio, 1920–1939 (Oxford: Oxford University Press, 2020).

39 Van der Vleuten, "Toward a Transnational History of Technology: Meanings, Promises, Pitfalls."

40 Michael Brüggemann and Hagen Schulz-Forberg, "Becoming Pan-European? Transnational Media and the European Public Sphere," *The International Communication Gazette* 71, no. 8 (2009): 693–712.

41 Ester Lo Biundo, *London Calling Italy: La Propaganda Di Radio Londra Nel 1943* (Milan: Unicopli, 2014); Nelson Ribeiro, "Censorship and Scarcity: Controlling New and Old Media in Portugal, 1936–1945," *Media History* 21, no. 1 (2015): 74–88, https://doi.org/10.1080/13688804.2014.950951; Stephanie Seul and Nelson Ribeiro, "Revisiting Transnational Broadcasting: The BBC's Foreign-Language Services during the Second World War," *Media History* 21, no. 4 (2015): 365–77.

ing the Brüggermann et al. categorization, there are also studies of transnational correlations between two or more types of media. For instance, Hilmes has traced the transnational relations between Britain and America that shaped the founding decisions with respect to broadcasting in both countries.[42] In this context, the approaches of transnational history and *histoire croisée* are synonymous because they are needed only to acknowledge that 'the very strength of the 'analytical cage' of a nation is a marker of the inherently transnational nature of the medium, and of the transnational cultural economy it creates.'[43]

The actors that shaped the development of radio were also operating transnationally. The national literature on radio development is frequently limited to national heroes who developed a notion of national radio broadcasting that served as 'an ideal symbol of national togetherness.'[44] By contrast, the transnational approach helps to identify actors who exchanged information about recent inventions, achievements, and policy decisions on wireless communication across the national territories and used the radio on a daily basis without necessarily laying claim to any revolutionary achievements in the national domain.

Transnational research concerns itself with media companies and their founders as key actors regulating cross-border exchanges which in turn determine national landscapes and influence national histories. In particular, the political economy of communication serves as a useful tool when carrying out such research.[45] As Mosco puts it, political economy is in this sphere 'the study of the social relations, particularly the power relations, that mutually constitute the production, distribution, and consumption of resources.'[46] The political economy of communication typically focuses on the evolution of communication and media that are produced and distributed by profit-seeking organizations in capitalist industries.[47] Most typically, from a transnational point of view, this type of research concerns the supranational corporations that have a direct impact on national markets. The recent studies of global media organizations, such as Google or Facebook, in different countries are the most characteristic example of this type of research, but there

---

42 Michele Hilmes, *Network Nations: A Transnational History of British and American Broadcasting* (New York, Abingdon: Routledge, 2012).

43 Ibid.

44 Michele Hilmes and Jason. Loviglio, *Radio Reader: Essays in the Cultural History of Radio* (London: Routledge, 2002), xi–xii.

45 Winseck and Pike, *Communication and Empire: Media, Markets, and Globalization, 1860–1930.*

46 Vincent Mosco, *The Political Economy of Communication* (London: SAGE Publications, 2014), 24.

47 Janet Wasko, "The Study of the Political Economy of the Media in the Twenty-First Century," *International Journal of Media & Cultural Politics* 10, no. 3 (2014): 261, https://doi.org/10.1386/macp.10.3.259.

are also many other companies that have had an impact on the media market on a global scale. As Negus showed, in the 1980s, the corporations responsible for famous music recordings acted as transnational actors that could not be identified with their national origins, a fact that led to much tension among individuals and groups in particular local and regional contexts.[48] For this research into radiotelegraphy at the beginning of the twentieth century, the study of the Marconi Company is one of the most relevant. Various aspects of the Marconi Company have been discussed through the personality of the founder of this international company,[49] or through its reluctance in the first two decades of the twentieth century to adapt to the emerging broadcasting function of radio technology.[50]

Another important focus of transnational media research is on international organizations as important actors shaping transnational flows of information, people, ideas, and concepts, with the International Telecommunication Union (ITU) considered here to be a key transnational actor in the arena of telecommunications at the European level.[51] The research looks at the interaction of the ITU with the networks of experts and users of radio through regulations, correspondence, and publications. An examination of the transnational understanding of radiotelegraphy and its transformation over the course of World War I reveals important details and facts in the history of the development of radio and helps to uncover the key features of the metamorphosis that wireless telegraphy went through to give birth to radio broadcasting. Thus, this book focuses on the links and interactions between different countries, which happened both in international arenas and outside of them.

The history of early wireless mostly emphasizes its heroic origins, the pioneers of radio, their successes, and failures, but rarely focuses on the institutional dimension of regulations and engineering. Some recent studies, however, have

---

48 Keith Negus, "Global Harmonies and Local Discords: Transnational Policies and Practices in the European Recording Industry," *European Journal of Communication* 8, no. 3 (1993): 295–316, https://doi.org/10.1177/0267323193008003003.

49 Marc Raboy, *Marconi: The Man Who Networked the World* (Oxford: Oxford University Press, 2016).

50 Gabriele Balbi, "Wireless's "Critical Flaw": The Marconi Company, Corporation Mentalities, and the Broadcasting Option," *Journalism & Mass Communication Quarterly* 1, no. 22 (2017): 1–22.

51 Alexander Badenoch, Andreas Fickers, and Christian Henrich-Franke, "Airy Curtains in the European Ether: Introduction," in *Airy Curtains in the European Ether. Broadcasting and the Cold War*, ed. Alexander Badenoch, Andreas Fickers, and Christian Henrich-Franke (Baden-Baden: Nomos, 2013); Simone Fari, Gabriele Balbi, and Giuseppe Richeri, *The Formative Years of the Telegraph Union* (Newcastle-Upon-Tyne: Cambridge Scholar Publishing, 2015); Andreas Fickers, "Visibly Audible: The Radio Dial as Mediating Interface," in *The Oxford Handbook of Sound Studies* (Oxford: Oxford University Press, 2012), 411–39.

begun to fill this gap by drawing attention to the significant influence and contributions of institutions. Recent studies in European media history have paid attention to such institutional actors as the ITU,[52] the International Broadcasting Union,[53] the European Broadcasting Union,[54] and various British institutions in the wireless industry.[55] These institutions were crucial gateways for transnational interaction. They mediated the shifting discourses about the role of broadcast media and established technical and/or juridical regulatory regimes that shaped media development.[56] Fickers has claimed that the organizations could be analyzed in terms of the concept of an 'arena': a site and space in the transnational mediascape where international discussions arose among such actors as media industry players and institutions.[57] In this research, the key arena to trace the development of wireless is the International Radiotelegraph Union, not only as an independent player, but also as a site that facilitated interaction between different actors and networks.

The transnational approach in war studies

World War I remained for a long time the more or less exclusive preserve of war studies, which by definition concerned itself with clashes between nations, and only recently has drawn attention to the mutual interferences that transcend a national framework. The national model was 'the easiest and most natural' as regards the task of describing and conducting research into the war, since nationalism was, in general, the driving force behind the conflict.[58] The national histories, 'written by the victors,' national museums, and the majority of war memorials embodied the idea of national difference and separation between 'us' and 'them.' The histories of the conflict written by German, British, French, and Russian historians

52 Gabriele Balbi et al., *Network Neutrality: Switzerland's Role in the Genesis of the Telegraph Union, 1855–1875* (Bern: Peter Lang, 2014).
53 Lommers, *Europe – on Air: Interwar Projects for Radio Broadcasting.*
54 R Zeller, *Die EBU – Union Européenne de Radio-Télévision (UER) – European Broadcasting Union (EBU). Internationale Rundfunkkooperation Im Wandel* (Baden-Baden: Nomos-Verlag, 1999).
55 Elizabeth Mary Bruton, "Beyond Marconi: The Roles of the Admiralty, the Post Office, and the Institution of Electrical Engineers in the Invention and Development of Wireless Communication up to 1908 (PhD Thesis)" (The University of Leeds, 2012).
56 Badenoch, Fickers, and Henrich-Franke, "Airy Curtains in the European Ether: Introduction."
57 Fickers, "Seeing the Familiar Strange: Some Reflections about Actants, Actors and Arenas of Transnational Media History."
58 Michael S. Neiberg, "Toward a Transnational History of World War I," *Canadian Military History* 17, no. 3 (2012): 31.

sometimes showed the event from opposite and even contradictory points of view. World War I was not exceptional in being dominated by national narratives.

However, war studies have themselves recently begun to adopt a transnational approach. A few studies have now represented the European world during the World War conflicts as a place where people of different nations shared much in common. Political views, religious ideals, social classes, everyday practices, and other phenomena transcended national borders. The constant exchange of information, concepts, and experts across national borders provoked the consolidation and separation of Europe into bigger forces than just nations. Moreover, the war stimulated constant transnational exchange: information, goods, experts, and people moved across borders to avoid or to support the conflict. The borders themselves were also continually changing.

World War I was an epochal event of in the European history, and some of its characteristics have been described from a transnational perspective. Neiberg drew attention to three important factors that reveal the transnational nature of World War I[59]: first of all, political movements spread in different nations, inspiring people from different contexts to engage in confrontations; second, the soldiers, enduring deplorable conditions in the trenches, shared the demands for fundamental economic and political change; and finally, a sense of disappointment was a common reaction in different nations to the post-war agreement.

Moreover, World War I is known as one of the first wars where technological progress played a crucial role worldwide. This technological progress shocked at different sensorial levels, from an all-pervasive fear of the onset of a gas attack to a completely transformed auditory landscape. The visual overstimulation in towns and cities with advertisements, buildings, traffic, and crowds described by Georg Simmel was accompanied by anxieties about rising noise levels.[60]. Shell-shock was an extreme form of this nervousness pervading the experience of modernity in the early twentieth century. Tanks were first invented in this war and proved to be an highly effective solution for massive attacks when produced in sufficient quantities. Military aviation also had its beginnings in this war; despite the primitive aircraft and self-trained pilots, countries nevertheless witnessed their first strategic air bombings. Warships had also seen great technical improvements, with naval mines being deployed in far higher numbers than in previous conflicts. Chemical weapons were first used systematically in this war as well, while machine guns likewise saw major use in World War I. Due to these techno-

59 Neiberg, "Toward a Transnational History of World War I."

60 James G Mansell, *The Age of Noise in Britain: Hearing Modernity* (Urbana, Chicago, and Springfield: University of Illinois Press, 2017).

logical advances, the idea of war was completely transformed, coming to rely more on technological equipment than on human force.[61]

Wireless telegraphy was one of the many technological innovations. War and military studies also address the issue of secrecy and transparency, which is a crucial topic for communication studies as well. The dichotomy of secrecy/transparency has a long history in research and goes back to Emmanuel Kant, Jacques Derrida, and Michel Foucault and the notion of the secret.[62] Despite the intention to sustain either transparency or secrecy, history is filled with the contradictions and interplay between these two terms.[63] The intention to transmit information transparently or secretly can serve to demonstrate the status of the sender or even be used in an opposite manner; transparency can play the role of security if the information used promotes a specific understanding of war and peace.[64] This opposition between secrecy and transparency finds its expression also in telecommunications and media history, as communication was always an 'invisible weapon' for nations and one that has played a crucial role in international politics.[65]

Transnational science and technology studies

Science and technology studies (STS) have also provided a necessary toolbox for this transnational research. STS focus on the relations between society, politics, and culture on the one hand and the development of technology and science on the other. Since radio was a technological innovation, most of the terminology and concepts of the STS approach were used for studying this medium. In general, STS and communication studies have much in common, particularly an under-

61 Stephen Bull and Adam Hook, *World War I Trench Warfare (1): 1914–16* (Oxford: Osprey Publishing, 2002); John Ellis, *Eye-Deep in Hell: Trench Warfare in World War I* (Baltimore: Johns Hopkins University Press, 1989); Spencer Tucker, *The Great War, 1914–18* (Bloomington: Indiana University Press, 1998).

62 Padideh Ala'i and Robert G Vaughn, eds., *Research Handbook on Transparency* (Cheltenham, Northampton: Edward Elgar, 2014); Suzanne J Piotrowski, ed., *Transparency and Secrecy: A Reader Linking Literature and Contemporary Debate* (Lanham, Boulder, New York, Toronto, Plymouth: Lexington Books, 2010).

63 Clare Birchall, "Introduction to "Secrecy and Transparency": The Politics of Opacity and Openness," *Theory, Culture & Society* 28, no. 8 (2011): 7–25, https://doi.org/10.1177/0263276411427744.

64 Dan Lindley, *Promoting Peace with Information: Transparency as a Tool of Security Regimes* (Princeton, Oxford: Princeton University Press, 2007).

65 Daniel R. Headrick, *The Invisible Weapon: Telecommunications and International Politics, 1851–1945* (New York, Oxford: Oxford University Press, 1991).

standing of the causal relationship between technology and society as one of 'mutual shaping' and an attention to their continuities and discontinuities.[66]

The most important principle of STS is that the invention of a technology is not an autonomous event that drags social change forward but instead is an inherently social phenomenon that is shaped by decisions on production, usage, and governance taken by particular people. A technological innovation is seen to be a consequence of a certain set of social, economic, and political circumstances.[67]

Within the field of STS, one of the most prominent theories of the social construction of technology (SCOT) draws upon social constructivism. According to this theory, human behavior directly shapes science and technology; therefore, their invention and development are 'not themselves natural.'[68] A detailed exploration of this line of enquiry may be found in studies by Pinch and Bijker, who have argued that the success of a technology depends on the power and scope of the groups that promote it.[69] The technology is co-constructed in debates, negotiations, and conflicts between and within the relevant social groups. These 'relevant social groups' impose different interpretations of technology according to their competing goals. These visions of technology compete in the initial stage of technology development, which is defined as the 'interpretative flexibility' of technology. Eventually, the predominant meaning of the technology in question emerges, and it undergoes stabilization and closure. Stabilization has to do with the character of the process, which could take several years, while closure is defined as 'the irreversible end point of a discordant process in which several artefacts existed next to each other.'[70]

The idea of relevant social groups developing in and through the social construction of technology has proved particularly useful for drawing attention to the users. In the linear model of technological diffusion, users are considered as merely passive consumers, whereas the social construction of technology acknowledges their capacity to transform the technology. Some scholars have also

---

66 Pablo Boczkowski and Leah A. Lievrouw, "Bridging STS and Communication Studies: Scholarship on Media and Information Technologies," in *The Handbook of Science and Technology Studies*, ed. E.J. Hackett et al. (Cambridge (MA), London: The MIT Press, 2008), 949–77.

67 Judy Wajcman, *Pressed for Time. The Acceleration of Life in Digital Capitalism* (Chicago and London: The University of Chicago Press, 2015), 4.

68 Sergio Sismondo, "Science and Technology Studies and an Engaged Program," in *The Handbook of Science and Technology Studies*, ed. E. Hackett et al. (Cambridge (MA), London, 2008), 14.

69 Trevor J. Pinch and Wiebe E. Bijker, "The Social Construction of Facts and Artefacts: Or How the Sociology of Science and the Sociology of Technology Might Benefit Each Other," *Social Studies of Science* 14, no. 3 (1984): 419–24.

70 Wiebe E. Bijker, "How Is Technology Made?-That Is the Question!", *Cambridge Journal of Economics* 34, no. 1 (2009): 69, https://doi.org/10.1093/cje/bep068.

shown that successful individual experiments sparked interested in radio at a mass level.[71] In particular, radio amateurs as an informal network of technical elites play a crucial role in adopting the technology into everyday life practices. The 'turn to users' in the scholarship has also prompted research on the intersection of feminist studies of technology and communication studies. Scholars had become aware of the need to go beyond stories about masculine practices of using the machines and demonstrated that women could also be considered active participants in technological change.[72] All these users played an important role in the 'domestication' of technology, whereby something unfamiliar and threatening was transformed into a familiar object embedded in everyday life practices and in the home environment.[73]

This research brings into focus curious amateurs, experimenters, and hobbyists. At the time, they were not yet considered as an institutionalized community. To grasp the informal nature of this activity, the combination of different theoretical concepts was used. First, scholars have deployed the notion of participatory culture.[74] Although usually applied to Internet projects like Wikipedia, YouTube, and Facebook, historians of technology remind us that this participatory culture is not a unique characteristic of our time. In *Selling the Air*, Streeter shows the network of early ham radio enthusiasts to have been a bottom-up, participatory culture that resulted in the development of modern broadcasting.[75] In fact, the first radio amateurs provoked the 'double birth' of media technology by adopting it to public needs and also re-inventing its use.[76] Second, we consider a radio amateur to be a person who experiences the pleasure of using the technology and is

---

71 Susan J. Douglas, *Inventing American Broadcasting, 1899–1922* (Baltimore, Maryland: The Johns Hopkins University Press, 1989).

72 E.g., Nina Wakeford, "Networking Women and Grrrls with Information/Communication Technologies: Surfing Tales of the World Wide Web," in *Processed Lives: Gender and Technology in Everyday Life*, ed. J. Terry and M. Calvert (London and New York: Routledge, 1997), 51–66.

73 Roger Silverstone and Erik Hirsch, eds., *Consuming Technologies: Media and Information in Domestic Spaces* (London: Routledge, 1992).

74 Henry Jenkins, *Textual Poachers: Television Fans and Participatory Culture* (New York, London: Routledge, 1992); Henry Jenkins, *Convergence Culture: Where Old and New Media Collide* (New York and London: New York University Press, 2006); Henry Jenkins, *Fans, Bloggers, and Gamers: Exploring Participatory Culture* (New York, London: NYU Press, 2006); Henry Jenkins, *Confronting the Challenges of Participatory Culture: Media Education for the 21st Century* (Cambridge (MA), London: MIT Press, 2009); Alvin Toffler, *The Third Wave* (London: Collins, 1980).

75 Thomas Streeter, *Selling the Air: A Critique of the Policy of Commercial Broadcasting in the United States* (Chicago, London: University of Chicago Press, 1996).

76 Gabriele Balbi and Simone Natale, "The Double Birth of Wireless: Italian Radio Amateurs and the Interpretative Flexibility of New Media," *Journal of Radio and Audio Media* 22, no. 1 (2015): 26–41, https://doi.org/10.1080/19376529.2015.1015860.

rewarded by his or her relationship to the object. For hobbyists, this technology was both a non-professional and a non-profit activity. The concept of enjoyment has recently become very much appreciated in the history of technology[77]; in *Enjoying Machines*, Brown and Juhlin[78] argue that enjoyment is the driving force in the proliferating use and accelerating development of technology, from gramophones to computer games. Indeed, radio amateurs, as is evident from their name, take a keen interest in and feel a deep affection for wireless technologies. Their usage of technology is inspired not only by great achievements in communication but also by the ludic aspect of technology.[79] Finally, in her analysis of technical hobbies, Haring argues that 'radio hobbyists formed their own 'technical culture,' a culture built around and establishing an ideology about technology.'[80] This technical culture provided the community of radio amateurs with the technological identity of a socio-technological group. This community has specific practices, ethical codes, and rules, such as 'social feeling, brotherly spirit, tolerance, politeness, and comprehension,'[81] which is usually obtained through the shared practices, the exchanges of knowledge, and the similarities in the patterns of using technology.

STS has also experienced the influence of a transnational approach. The transnational history of technology focuses on networks that operate across national borders and highlights actors that initiated and endorsed the development of these projects.[82] In the late 1990s, scholars started studying European history through the lens of technology and discovered that major changes overtook Europe in many different national landscapes. An important actor within this academic field was the research network 'Tensions of Europe', which embodies the idea of a transnational perspective on the development of technology in Europe by establishing correlations and links between infrastructures, experts, goods,

77 Alfie Bown, *Enjoying It: Candy Crush and Capitalism* (John Hunt Publishing, 2015).

78 *Enjoying Machines* (Cambridge (MA), London: The MIT Press, 2015).

79 Alfred Kirpal, "Ernst Oder Spiel? Basteln, Konstruieren Und Erfinden in Der Radioentwicklung," in *Homo Faber Ludens. Geschichten Zu Wechselbeziehungen von Technik Und Spiel*, vol. 4 (Frankfurt am Main: Peter Lang, 2003), 227–54.

80 Kristen Haring, *Ham Radio's Technical Culture* (Cambridge (MA), London: The MIT Press, 2007), xv.

81 Francesco Schiavone, *Communities of Practice and Vintage Innovation: A Strategic Reaction to Technological Change* (Heidelberg, New York, Dordrecht, London: Springer Science & Business Media, 2014), 93.

82 Erik van der Vleuten and Arne Kaijser, "Networking Europe," *History and Technology* 21, no. 1 (2005): 21–48, https://doi.org/10.1080/07341510500037495.

knowledge, and resources.[83] The key concept here is that of transnational infrastructures, from the railway to the telegraph, which conjoin and unite the efforts of different countries in constructing efficient and cohesive networks.

In its untidiness and uneven spread, infrastructure is an important concept for the history of technology because the physical type of network is analyzed through its construction, linking, and maintenance. Technological infrastructures form the physical basis for transnational flows of people, goods, and services and represent 'the essence of European integration.'[84] To construct a coherent infrastructure, empires and social groups expanded and re-negotiated transnational relations. Infrastructures also provided new ways of experiencing Europe in its 'hidden integration' and co-operation, such as through transportation or communication. The concept of 'hidden integration' refers to the construction and use of transnational infrastructures that introduce an understanding of Europe as a coherent, united space. Schipper and Schot refer to this process of constructing Europe through infrastructures as 'infrastructural Europeanism.'[85]

Archives and sources

Preservation of historical sources was historically a national prerogative; therefore, national archives were frequently the most common source for historical data. Transnational research, in contrast, necessitates the collection of archival sources that allow tracing of the flows and phenomena that exceed national frameworks. When undertaking transnational research, a scholar is obliged to collect and combine different types of sources coming from various national frameworks. As Douglas has noted, this leads to the creation of one's own archive. In other words, scholars set out to capture the media zeitgeist of the past, for only through this can they hope to gain some sense of the media environment, although even then the picture remains incomplete.[86]

---

83 Van der Vleuten, "Toward a Transnational History of Technology: Meanings, Promises, Pitfalls."

84 Badenoch and Fickers, "Introduction: Europe Materializing? Toward a Transnational History of European Infrastructures," 2.

85 Frank Schipper and Johan Schot, "Infrastructural Europeanism, or the Project of Building Europe on Infrastructures: An Introduction," *History and Technology* 27, no. 3 (2011): 245–64, https://doi.org/10.1080/07341512.2011.604166.

86 Susan J. Douglas, "Writing from the Archive: Creating Your Own," *Communication Review* 13, no. 1 (2010): 5–14, https://doi.org/10.1080/10714420903558613.

The book is based upon archival research into materials held at the International Telecommunication Union Library and Archives, Geneva; the Marconi Archives, Bodleian Library, Oxford; the archives of the Institution of Engineering and Technology, London; the British Telecom Archives, London; the Russian State Library, Moscow; the British Library, London; Dokumentationsarchiv Funk, Vienna; and PTT archive, Bern. The archival materials of other international organizations also shed light on the direct and indirect regulation of radio, such as International Time Conference and Congress of Geography, the International Broadcasting Union, and International Radio Amateur Union.

This book can be distinguished by its centralized organization of the archival research. It centers around the role and networks of the International Telecommunication Union, one of the most important international organizations as regards communications. The ITU is currently located in Geneva, Switzerland, and is an agency of the United Nations. At the beginning of the twentieth century, it was called the Telegraph Union and was based in Bern, Switzerland. It was created in 1865 to regulate and unite the national networks of electric telegraph so that information could more easily cross borders,[87] and was the very first international organization, predating even the Red Cross. It was one of the most important actors in the telecommunication industry and in the 1900s was entrusted with the task of taking on a leading role in regulating radio. The ITU was an important arena where the interaction between various actors occurred: not only at international conferences, but also in information exchange and correspondence beyond international events.

In order to trace the network of actors involved in the discussions around radio, an extensive collection of archival documents has been studied, which comprises a combination of different sources on an international and a national level. The archival documents and literature are in English, French, German, Italian, Spanish, Portuguese, and Russian; the object of the study has therefore been researched in its different national and cultural contexts. All translations are my own, unless otherwise specified.

The first category of documents is the sources of the ITU Library and Archives in Geneva. Its archive shaped the research and revealed the core issues of the discussions about regulations on an international level. It represented a large-scale overview of radio development and offered an understanding of general trends, preserving as it did materials from governments, corporations, experts, and users on problems of transnational importance. Its main collection includes

87 Balbi et al., *Network Neutrality: Switzerland's Role in the Genesis of the Telegraph Union, 1855–1875.*

both published and unpublished documents. The latter were produced for members' use and usually served as a basis for the organization of the Bureau. The vast majority of the documents are in French, which was the official language of the Union, with some other internal documents written in German.[88]

The second category of archival documents comprises national sources preserved in numerous national archives and libraries. The selection of the sources was undertaken with a view to identifying the issues mentioned in the ITU documents and to tracing how they were discussed in different national settings, with this work also done with the help of secondary literature. As radio communication was predominantly Eurocentric, several countries were included in the scope of this research: the UK, Germany, France, Russia, Switzerland, Italy, and the US. The selection of these countries was driven by a research focus on Europe and included other non-European countries (such as the US) insofar as they were present or took an active part in discussions on radio in the ITU arena. It must also be underscored that these national archives focused not only on national issues but also frequently preserved documents from other countries. For instance, the Marconi archive at the Bodleian Library in Oxford contains the Russian journal *Vestnik telegrafii bez provodov* (transl. 'Herald of wireless telegraphy'), whereas in the Russian archives and libraries it was destroyed or damaged in the course of the October revolution and its aftermath in the Soviet Union.

The third type of archive had to do with sources to be found in international organizations that also dealt with the direct and indirect regulation of radio which were identified using documents of the ITU and also national documents. These documents were particularly crucial for responding to research questions about broadcasting's function and the transnational dimension of radio. The documents of the International Time Conference and Congress of Geography helped to address the broadcasting functions of the radiotelegraph, while the documents of the International Broadcasting Union and International Radio Amateur Union were relevant for tracing the transformation of radio after World War I.

Finally, to grasp the essence of this international landscape, the use of technology in everyday life was accounted for through notes, letters, and articles published in radio amateurs' reviews and magazines. These sources were particularly useful because they contained not only radio amateurs' comments and observations but also commented on official reports from the international meetings. Moreover,

88 The documents included International conventions on radiotelegraph (1912 and 1927) and telegraph communications (1925), Minutes of the international conferences of 1912, 1925 and 1927, ITU monthly periodical Journal télégraphique (187 issues), ITU annual reports, Historical statistics, Radiotelegraphy maps (1922/1923; 1925/1926; 1927), Correspondence of the ITU and its register (34,954 items).

sources relevant for this timeframe were also collected from the archive on radio amateurs in Vienna, the Dokumentationsarchiv Funk, which preserves documents directly from radio amateurs and has a vast collection of QSL-cards.

Apart from the use of voluminous primary sources, the novelty of this research is that it has brought three underestimated types of archival documents into the analysis and revealed a particular pattern in the preservation of archival sources of transnational importance today.

First of all, this research offers a pioneering analysis of radiotelegraphy maps. Most often maps are used only as an illustration of the collected data,[89] with rare exceptions of in-depth analysis.[90] However, critical research into maps, primarily based on the work of Harley,[91] has led to the reinvention of the map as a politically engaged document with a meaningful impact on society. As historians of technologies have established, maps represent a crucial source of information, by virtue of their forming a part of a network. They maintain the status quo of this network because they imply the idea of the sustainability of the given network and depict its cohesion. They are also a part of the 'reality' of these infrastructures; for example, one would never take a train from Manchester Airport to Leeds if it were not depicted on the map or one would never send a telegram to Barcelona if there were no channel of communication depicted or announced somewhere. Maps create myths.[92] This monograph seeks to explore maps as a type of social construction of knowledge. Phenomena that appear as neutral technical practices on maps often mask their complex relations and political engagement. To extract different layers of meaning from a map, one must also see the silences and margins. Maps select, articulate, frame, structure, bind, and name objects in physical reality in the process of transforming them into a two-dimensional image. Even 'simple' naming has a political significance on maps. For this particular research, maps must be viewed not as an illustration of the existing networks but instead as a representation of the particular selection of artefacts, trends, and contexts relevant for radio communication.

Second, another important discovery throughout this research has concerned the visual dimension of 'invisible' technology. As previously mentioned, maps pro-

89 Helgi Björnsson, "Radio-Echo Sounding Maps of Storglaciären, Isfallsglaciären and Rabots Glaciär, Northern Sweden," *Geografiska Annaler. Series A, Physical Geography* 63, no. 3/4 (1981): 225.

90 Lisa Parks, "Earth Observation and Signal Territories: Studying U.S. Broadcast Infrastructure through Historical Network Maps, Google Earth, and Fieldwork," *Canadian Journal of Communication* 38 (2013): 285–307.

91 J.B. Harley, "Deconstructing the Map," *Cartographica: The International Journal for Geographic Information and Geovisualization* 26, no. 2 (1989): 1–20.

92 J.B. Harley, "Maps, Knowledge, and Power," in *The Iconography of Landscape: Essays on the Symbolic Representation, Design and Use of Past Environments* (Cambridge: Cambridge University Press, 1988), 277–313.

vided a source for understanding radiotelegraphic networks and their construction. In addition to maps, cartoons, drawings, and photographs also presented a valuable historical resource. Their images present a visualization of the common beliefs, feelings, and anxieties regarding the new technology, dealing directly with how people imagine media. Radio technology is assigned a very different role in society if it is depicted with lines, circles, dots, or indeed with anything else.

Third, particular attention should also be paid to a previously overlooked kind of archival document: the hand-written correspondence register, which represents another contribution to media history (Figure 2). This correspondence register is focused on radio communication only; it is an internal document that preserves information about each letter or notice sent or received by the ITU, with the name of the sender or addresser (or addressee), date, city, and a brief explanation of the content. Researchers often overlook the correspondence register as a historical source, using it instead as an index to find specific documents

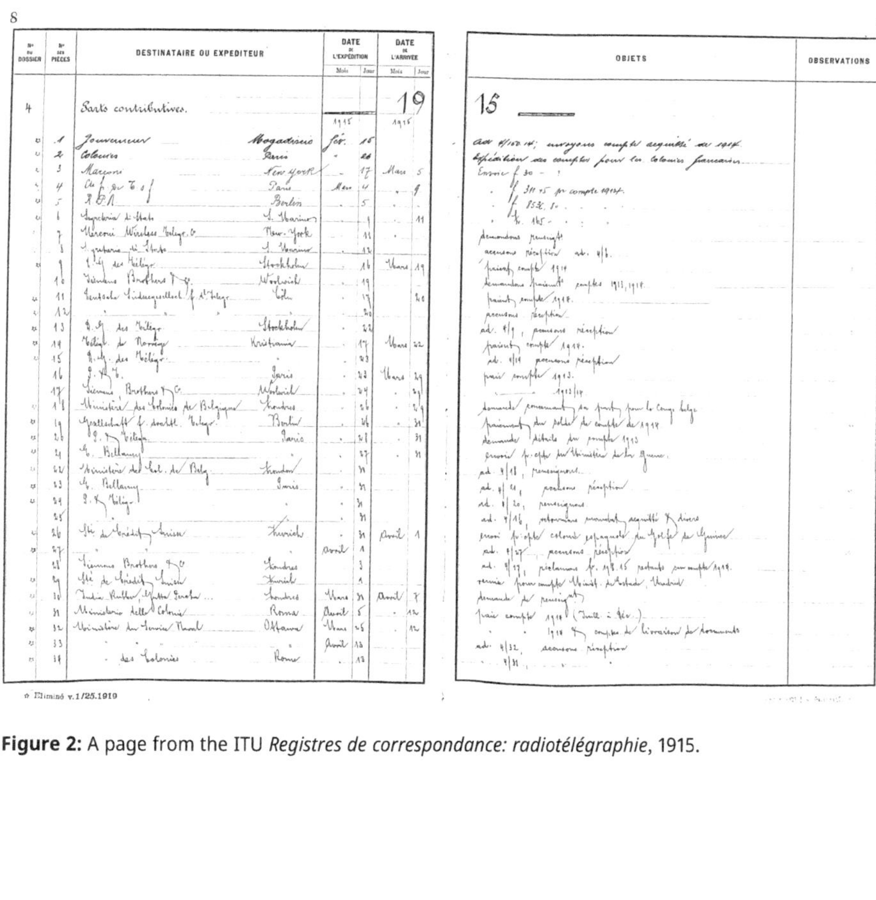

Figure 2: A page from the ITU *Registres de correspondance: radiotélégraphie*, 1915.

and turning it into a proper reference only if the original file were destroyed.[93] Correspondence registers have become an object of research mainly in studies of record keeping, especially in comparison with digital systems.[94] Scholars have acknowledged that analogue books represent an important source of information as they offer more freedom in filling in gaps and adding supplementary information between the lines, in contrast to the digital forms.[95] In other words, the correspondence register is a marginalized source for the media historian but may prove useful precisely because it preserves comments in the margins.

In this research, an analysis of the ITU correspondence register helped to understand which countries, institutions, and people were interested in the development of radio, who followed the news and who was most closely connected to the ITU. Figure 2 is an example of a typical page from the ITU correspondence register on radio in 1915. As the Telegraph and Radiotelegraph Union were seen as two separate focuses of the ITU Bureau, correspondence with the state-members and other actors regarding these topics were also preserved separately. The radio correspondence register is concerned only with matters regarding the Radiotelegraph Union. The hand-written table is organized in several columns, keeping information about the sender or addresser of the message, the city of origin, the date sent and received, and the content of the message (cursorily described) with reference to others if it was a mail thread. At first, all letters were preserved, and these registers indeed were used as an index for these records. However, over time, they also became more useful as they contained information from the archive that had already been destroyed. In the column furthest to the left, archivists marked line with a star if the original item had been destroyed, which may have occurred due to extraordinary circumstances, such as World War I or a fire, or perhaps the reorganization of the archive. The page reproduced in Figure 2 indicates that the items were marked as destroyed on January 25, 1919, and a reader may see from the image itself that very few original items were left in the archives: only five out of 34, which therefore turns this register into a unique and indispensable source of information. The existence of such a document allowed the obstacle of having some materials destroyed to be overcome.

93 Marika Sherwood, "Strikes! African Seamen, Elder Dempster and the Government 1940–42," *Immigrants & Minorities: Historical Studies in Ethnicity, Migration and Diaspora* 13, no. 2–3 (1994): 130–45.

94 Caroline Allinson, "The Process of Audit and Control – A Comparison of Manual and Electronic Information Systems," *Policing: An International Journal of Police Strategies & Management* 27, no. 2 (2004): 183–205, https://doi.org/10.1108/13639510410536814.

95 Gudmund Valderhaug, "Recordkeeping in Local Government in Norway 1950 – 2000," *Archival Science* 3 (2003): 205–12.

As correspondence with many actors was regular and consistent, this register represents a unique database of the network organized around the ITU regarding radio communication, which also helped in further research at national archives and libraries and allowed a reconstruction of the knowledge infrastructure around the ITU. Many contributions of different state-members, reports of commercial companies, and requests from private individuals could be found. Since the ITU was the most important and influential international organization in regulating radio communications and was accumulating national data at the beginning of the twentieth century, correspondence about radio was often addressed to the Bureau with a view to obtaining recent information about the development of the technology. For instance, many letters included enquiries about statistics, circulars, and recent issues of the official magazine. Sixteen correspondence books were chosen for analysis: from 1911, the first year when the correspondence for radio was filed separately by the ITU, to 1927, the year of the Washington conference that dealt with the consequences of World War I. The overall number of unique pieces (messages, letters and invoices) sent and received during these years by the ITU is 34,954. The analysis of this register was carried out in several different stages: transcription, checking the actors, analyzing particular pieces of correspondence, and clarifying the details with the use of additional sources from *Journal Télégraphique* and other historical documents. The understanding of what kinds of actors featured most frequently in communications with the ITU and what topics they were raising most often adds a fresh perspective to the transnational picture of radio development.

There were also some particular obstacles to conducting the archival research that had to be overcome. The closure of the International Broadcasting Union, an important archive that is no longer accessible to researchers and the general public, was a particular loss for this historical research but, fortunately, the absence of its documents was compensated for by archival materials of the ITU. It was discovered that the two organizations did exchange protocols of their meetings, and the ITU Archives preserved many of these same documents.

This archival research also led to a surprising discovery. Just as in the early twentieth century, when different actors, institutions, and organizations exchanged information, today their archives constitute a particular network that replicates the same pattern of communication. The ITU archives turned out to be a location where documents were consolidated not only because the ITU used to be a crucial organization but also because of its success in continuing to be one of the key international archives on telecommunications until the present day. In particular, it was discovered that the British Telecom Archive has also recently sent some of its historical material to the ITU in Geneva. Similarly, the BT and IET archives also recently exchanged some collections to construct a complete picture of their history.

Therefore, the archives on the same topic today replicate the network of actors in the twentieth century, and the materials today travell all the way back to their places of origin.

Limitations of the present study

First of all, this research is limited to the ITU's network, therefore leaving some important national actors outside of the framework. In particular, some important European countries were included in the research only to the extent of their participation in international discussions.

Then, this research is largely based on British documents and involves many British sources in contrast to other countries, which could be considered a big limitation or an analysis of the transnational influence of European radiotelegraphy. This trend is, however, justified for two reasons. First, Britain indeed was the most dominant country in Europe in the development of radiotelegraphy, which could easily be seen from the number of radiotelegraph stations (see Figure 7 and Appendix A). Second, Britain was in extensive communication with other countries and collected and preserved documents from all over the globe. As already mentioned, the Russian State Library only has a few issues of the Russian Herald on Wireless Telegraph, as many were destroyed or lost over the course of the October Revolution, while the Bodleian Library in Oxford actually holds all issues of this journal. Therefore, this dominance of British archives allowed us to draw meaningful conclusions on the state of radiotelegraphy in Europe in general.

The technical language of some experts' reports also constituted a problem, especially when physicists' reports on experiments, mathematical explanations, and formulas did not consistently use the same terms, particularly in regards to their translations. This problem of understanding and interpreting the technical sources was, however, solved through consultations with physicists and engineers.

Finally, the research encountered some challenges with the absence of specific primary sources, which had to be substituted with other sources. For instance, while some actual correspondence had been eliminated and was therefore no longer preserved in the ITU Library and Archives, its correspondence register served as the most important source because it preserved information about the brief content of the missing letters. Another significant lack of information regards the closure of the IBU archive to researchers, however, its documents were replaced with communications found in the ITU Library and Archives collection, as these two institutions had once exchanged the most valuable protocols. Moreover, the most recurrent documents appear numerous times in different national archives, which leads to a historiographical reflection; transnational research has some advantages in respect to any other historical research, as the primary sources absent in one archive could possibly be found in another. Just as the most crucial telecommunication actors of the early twentieth century formed a network and exchanged information, today their archives also trade information and form a kind of network.

As previously mentioned, one of the limits of this study was its focus on the transnational networks of actors around the ITU, which also influenced the Eurocentric character of this work. In future studies, the inclusion of the histories of radio in other countries would help to achieve a more coherent picture of radio's development and shed more light on its transformations. In particular, some European countries on the margins of the ITU network were only briefly presented or even left out of the picture, especially those in northern and south-eastern Europe. Information from other parts of the world would also contribute greatly to understanding transnational networks, especially South America and Asia due to increasing interest in them and their influence during World War I and afterwards. In other words, to better trace the transformations and development of transnational networks, it would be beneficial to go beyond the ITU networks of actors in further research.

Chapter outline

The order of the chapters reflects the chronological development of wireless telegraphy and radio broadcasting. The chapters proceed from the early 1910s to the late 1920s, however, occasionally, the rules of strict chronology are broken in the interests of the overall narrative. Apart from the first chapter on the origins of wireless and its controversial character, chapters 2–4 have a similar structure. They trace the challenges faced by, and the changes to, the international arena of wireless telegraphy, the political economy of radio communication, transnational radiotelegraph networks, and the role of radio amateurs. The fifth chapter serves as a conclusion, highlighting the most important lessons to be drawn from wireless development throughout the twentieth century.

Chapter 1, *The Innovative and Controversial Wireless*, provides an overview of the origins of wireless telegraphy in the 1900s. First of all, it sets out to identify what wireless is, discusses its use for point-to-point and one-to-many communication, and considers various controversial and contested names that were used interchangeably for wireless: wireless telegraphy, radiotelegraphy, radiotelephony, and radio broadcasting. The controversy surrounding the names and applications reveals the 'interpretative flexibility' of the technology,[96] as it was open to a range of different interpretations and experiments. The competition between interpretations and experiments resulted in a solid 'closure' or even a 'double birth' of the technology in the form of the national radio broadcasting services in the early 1920s. The chapter then goes on to map out the trajectory of the wireless industry, from the patent wars to the peculiarities and advantages of different key companies, such as the non-intercommunication policy adopted by the Marconi Company and the advanced ground communication (in contrast to maritime telegraphy) espoused by Telefunken. Furthermore, as wireless easily transcended borders and was used for communication at sea as well as across international borders, the chapter accounts for the formation of the International Radiotelegraph Union (established alongside the International Telegraph Union). Finally, the chapter describes the most critical uses of the technology and of the users of the technology: its deployment at sea, for private communication, and social fears surrounding connection without cables.

Having set the scene for wireless industry and policy in the 1900s, the book proceeds with Chapter 2, *Wireless for a Global Audience, the Early 1910s*, which features an unusual take on wireless history, focusing on global networks and connectivity in the 1910s. The chapter begins with the history of the Titanic tragedy, which represented a turning point in the history of wireless, demonstrating that wireless could have saved more passengers had the requisite transnational collaboration already been in place. By analyzing the ITU 1912 Conference materials, the chapter demonstrates how the new set of international regulations on wireless shaped a transnational understanding of radio communication and required collaboration beyond national boundaries. It further explores the political economy of radio communication and reveals how the competitive wireless industry consisted of 'born global' companies that had to operate transnationally. It takes the political economy of communication to consist of 'the power relations, that mutually constitute the production, distribution, and consumption of resources.'[97] The chapter then goes on to discuss the global projects that were launched in the 1910s, among them the many attempts made, in light of the Titanic disaster, to improve safety at sea. The international time signals network was set up to transmit a uniform time signal through the combined efforts of ten radio stations around the globe. This project helped to achieve homogeneity and universality of time across space, but served also to create a transnational radio audience keen

96 Pinch and Bijker, "The Social Construction of Facts and Artefacts: Or How the Sociology of Science and the Sociology of Technology Might Benefit Each Other."
97 Mosco, *The Political Economy of Communication*, 24.

to receive signals from Paris even when in, say, London. Furthermore, the international meteorological network transmitted weather reports by radio waves across national borders, and ever since this has been one of the chief functions of radio. Moreover, experiments with voice and music broadcasting were rendered possible through the transnational reach of radiotelegraphy. Belgian radio concerts were, for example, regularly received in France. The chapter also discusses a project to design a universal language that would permit communication notwithstanding the existence of different national languages. There were thus experiments with Esperanto, and a set of abbreviations (Q-codes) were created with this transnational purpose in mind. Finally, the chapter reveals that the first users of wireless telegraphy, both the professional radio operators and the radio hams, constituted an informal network of technical elites that played a crucial role in adopting the technology into everyday life practices.[98] Moreover, it pays particular attention to the rapid and far-reaching growth of the wireless industry in the 1910s due in large part to the abovementioned projects, which led to the renegotiation of gender roles in the job market globally.

Chapter 3, *European Wireless at War, 1914–1918,* focuses on the challenges to wireless telegraphy posed by global war. The chapter surveys new applications of wireless, but also abandoned or discredited practices. First, it considers new applications of wireless telegraphy introduced or developed during World War I and its use by the military for communication between allies. For example, it addresses the inevitable transnational technology and knowledge exchanges that occurred, with many companies having branches in different countries and being nationalized with the outbreak of the war (such as, for instance, the inventions of German company Siemens, there being an office of the Siemens Brothers Company in Woolwich, London). Furthermore, it discusses the progress made in encrypting and coding by the Allies and Central Powers alike (especially in the case of the radiotelegraphy used by spies), the enhanced portability of the devices in question, and the use of voice in communication (essential as it was in aviation). It also discusses how radio began to be used by new groups of users, such as soldiers, children, and women. The chapter then goes on to draw attention to the ways in which the international arena changed with the outbreak of war, noting how conferences were deferred but also how the work of the ITU Bureau contin-

98 Balbi and Natale, "The Double Birth of Wireless: Italian Radio Amateurs and the Interpretative Flexibility of New Media"; Douglas, *Inventing American Broadcasting, 1899–1922*; Haring, *Ham Radio's Technical Culture*; Stephen Lippmann, "Boys to Men: Age, Identity, and the Legitimation of Amateur Wireless in the United States, 1909–1927", *Journal of Broadcasting and Electronic Media* 54, no. 4 (2010): 657–74, https://doi.org/10.1080/08838151.2010.519812; Streeter, *Selling the Air: A Critique of the Policy of Commercial Broadcasting in the United States.*

ued beyond the war, hence with non-European countries. The chapter further considers the restrictions placed on civil wireless telegraphy and transnational networks and the challenges that they faced, which resulted in the destruction of global radio infrastructure. Overall, it demonstrates how the war gradually destroyed the transnational infrastructures of radio, in effect the global media system of the 1910s, the destruction of which could also be seen as a destruction of 'the essence of European integration.'[99] Finally, the chapter demonstrates how new ways of using radio that were national in scope gradually appeared: the boom in national wireless news agencies during the war, the suspension of radio amateurism in Europe, and the activities of radio hams being declared illegal (in contrast to the case of their American counterparts).

In Chapter 4, *Establishing a New Order, 1918–1927*, the aftermath of World War I is discussed. First, the chapter recounts the initial, tentative attempts to restore the global networks, the struggles in the international arena, the attempts of the ITU to gather information, the formation of the new international organizations and the ITU 1927 Conference in Washington, which forged a new understanding of radio in terms of a new international convention. The chapter then covers the launch of radio broadcasting stations and the transformation of the radio sector into the broadcasting industry. This chapter also traces the transformation of the transnational radiotelegraph networks by considering the case of weather reports and time signals as well as radiotelegraph maps. Finally, it traces the institutionalization of radio amateurism in Europe and the formation of national organizations responsible for managing the international amateur service and containing all transnational communication within the bounds of national legislation. Overall, this chapter explains how the wholesale destruction of the transnational networks, infrastructure, and communication during the war gave rise to radio broadcasting as a national medium.

The concluding Chapter 5, *The Global Wireless: Past and Present*, describes the contributions of this monograph to our understanding of wireless and radio history. It traces wireless development throughout the twentieth and into the twenty-first century, from telegraphy to Wi-Fi and 5G networks, and discusses the most important legacies of wireless telegraphy. The chapter considers various different aspects of global communication networking by wireless: the wireless transmission of electricity, from Tesla's ideas to IKEA wireless chargers; the integration of wireless with cable infrastructure and the similarities between integration of wireless and wireless telegraph with mobile antennas to hyperoptic cables; the issue of security as it

99 Badenoch and Fickers, "Introduction: Europe Materializing? Toward a Transnational History of European Infrastructures," 2.

relates to radio waves and the imaginary of wireless as a secure channel from the Marconi presentations to Wi-Fi standardization; small scale wireless networks among amateurs then and now; and the persisting dilemmas regarding the radio spectrum, spectrum holes, licenses and other critical issues. For quite some time, the term 'wireless' was buried under the popularity of radio broadcasting; today, however, scholars emphasize that wireless technologies embed a particular 'wirelessness' that refers to a wider and deeper belief in media-technological progress.[100]

100 Adrian Mackenzie, *Wirelessness. Radical Empirism in Network Cultures* (Cambridge (MA): The MIT Press, 2010).

Chapter 1
The innovative and controversial wireless

> It was not the public that waited for radio but radio that waited for a public (. . .) It was suddenly possible to say everything to everybody but, thinking about it, there was nothing to say. Bertolt Brecht, 1932[1]

Wireless telegraphy was invented in the late nineteenth century and is primarily associated with the names of Guglielmo Marconi, Ferdinand Brown, Alexander Popov, and others. These inventors were among the first to design radio transmitters and receivers to send a signal on electromagnetic waves. They relied on the longstanding work of the many theoretical physicists, Michael Faraday, James Clark Maxwell, and Heinrich Hertz among them, who since the 1860s had studied radiation and waves, and had also made practical suggestions about their control and use for communication purposes, as well as for other things.

Throughout the twentieth century, wireless entangled the world with radio waves. Wireless allowed for communication over short or long distances without any cables, thereby rendering communication much easier and cheaper. Starting from Morse communication between ships and airplanes, it opened a new chapter in the world history of wireless communication: radio broadcasting, television, mobile networks, satellite communication, Wi-Fi, and 5G. The length and format of signal, the methods of encryption, and the ways of using wireless evolved, but the idea and the technique always remained the same – wireless communication sent between two or more antennas into the air by radio waves.

This chapter introduces the wireless as a technological innovation of the twentieth century. It discusses the invention of wireless, its main uses in the 1900s, and the key actors that shaped its development. It places wireless technology in the media environment of the day and demonstrates the challenges that technology encountered and the crucial decisions made for its development from the very start.

What is wireless?

A wireless telegraph is a system designed to transmit information by radio waves using Morse code or some other simple code. The spark system of wireless telegraph was designed as follows. The telegraph operator operates the radio transmitter with

1 Bertolt Brecht, "Radio as a Means of Communication. A Talk on the Function of Radio," *Screen* 20, no. 3–4 (1979): 24.

Open Access. © 2024 the author(s), published by De Gruyter. (cc) BY-NC-ND This work is licensed under the Creative Commons Attribution-NonCommercial-NoDerivatives 4.0 International License.
https://doi.org/10.1515/9783111202327-002

two knobs, one of which is connected to the aerial and the other to the wire running down into the earth. The device has a key, which the operator has to tap. By tapping the key, two knobs are charged with electrical energy, until the air between them is unable to keep it apart and produces a bright spark between them. This sends a shock both along the wire down into the earth and into the air through the antenna. The operator can tap short or long, creating two types of signals: a short 'dot' or a long 'dash.' The message arrives at the receiving office through its antenna, and the operator hears the sparks as a sequence of short or long 'beeps,' thus enabling the deciphering of the message. Typically, wireless telegraphers would transmit text information, but they could also encode any other kind of information. A series of short and long signals is nothing but a binary code that can be used to communicate any information, from texts to images and to music.

In the early days of radio communication, the wireless telegraph was the only way to transmit information over the radio waves. Even today, despite the emergence of new, much more effective modes of communication, the wireless telegraph is still used by amateur enthusiasts and sometimes in official communications.

The nature of wireless was absolutely ground-breaking: invisible, not tactile, and thus unimaginable, yet very efficient. However, as with any innovation, these revolutionary features of wireless were at first difficult to understand. People were literally scared of this new technology; just as today, when thousands are worried about the 5G antennas' impact, 100 years ago people were frightened to see how the message could travel without any physical connection. They associated wireless communication with mind-reading, telepathy, supernatural powers, with the wireless operators seen as nothing less than mediators with the netherworld. A sketch from the 1910s even pictures a ghost close to the operator, tapping the key (Figure 3).

Science and technology studies suggest that this skeptical and even troubled perception of a new technology is quite common. It is normal for people to have a social fear of something new, and therefore it is difficult for a technology to revolutionize social practices at once. Innovation-driven growth may be one of the most influential business ideas today, but a disruptive innovation[2] always arouses a great deal of skepticism in its opponents. One very common way for inventors to quell such opposition is to relate an innovation to something already known and widespread, in other words, to imitate an older medium.[3] By building continuity with the old technology and social practices, the new one manages to re-

2 Clayton M. Christensen, *The Innovator's Dilemma: When New Technologies Cause Great Firms to Fail, The Innovator's Dilemma* (Boston, MA: Harvard Business Review Press, 1997).

3 Gabriele Balbi, "Old and New Media. Theorizing Their Relationships in Media Historiography," in *Theorien Des Medienwandels*, ed. Susanne Kinnebrock, Christian Schwarzenegger, and Thomas Birkner Hrsg (Köln: Herbert von Halem Verlag, 2015), 231–49.

Figure 3: Ghosts accompany wireless telegraph operators at their work.[4]

duce the natural social fear and at the same time reveal the differences between them. That was the reason why for around the first decade, the application of wireless telegraphy was not so very different from existing communications, namely the electric telegraph and wired telephone. It was imagined similarly, and its innovative and disruptive characteristics were at first downplayed.

Being perceived as a cable substitute, wireless misleadingly acquired certain characteristics of the physical wire. There were two main controversial points regarding the nature of wireless, which, as time passed, would prove to be its greatest advantage, but which at first were simply too innovative and difficult to understand. Wireless radiated information over radio waves, yet was perceived as a point-to-point channel; wireless sent messages into the air openly, yet was perceived as a private and secure communication analogous to cable. The following paragraphs will serve to explain these points in depth.

First of all, even though wireless communication radiated in different directions over radio waves, for a long time it was imagined as a point-to-point channel. Its development was on a par with cable electric telegraphy, operated by the same departments, and often directly linked to the same network. The infrastructure of the electric telegraph and of the wireless telegraph were in fact quite similar in the early days. Telegraph offices had no problem adding a wireless option to their services; they installed antennas in their offices and trained personnel to tap a different key, which after all was not so very different: still a key and using the same Morse

4 "The Young Man's Eyes Fell on His Own Transmitter Key. It Was Clicking up and Down," *The Wireless World* 1 (1915): 639.

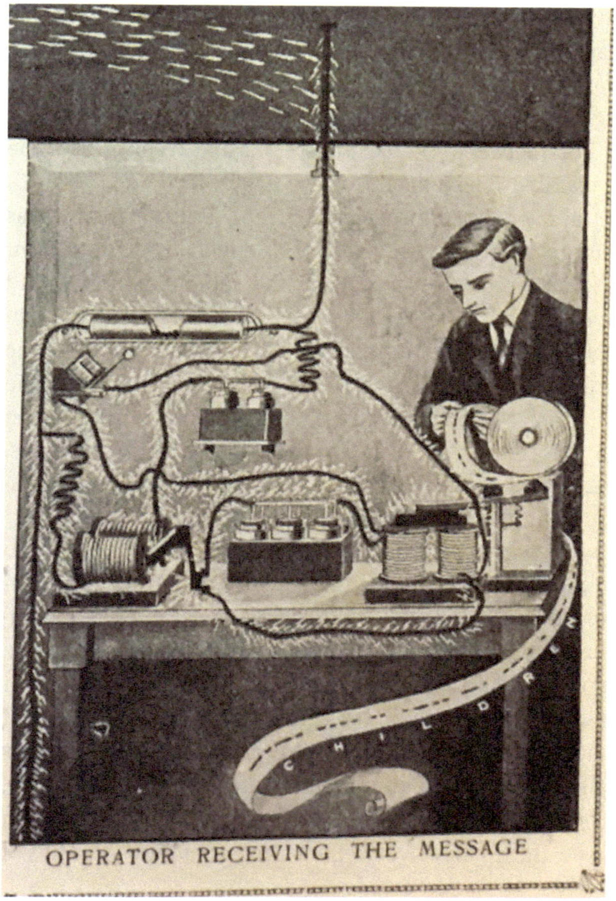

Figure 4: The shaded lines of wireless communication.[5]

code. Even though wireless did not use any cables, the network was still designed as if a cable existed. Wireless was perceived as a point-to-point medium, which meant that the wireless industry denied outright that the waves radiated and spread around. This means that wireless was at first presented as a channel between two respondents, and not as a transmission from a broadcaster to a large audience. Any

5 Arthur Mee, ed., "The Unseen Telegraph Messenger Turning Electric Waves into Words," in *The Children's Encyclopaedia* (Lewisham, London: Educational Book Company, 1914), 3526–27 (Fragment).

evidence of a broadcasting function was seen as a 'critical flaw';[6] the wireless was supposed to serve as a replacement to wired communication, thus providing a secure channel of communication, instead of uncontrollably disseminating information around. The explanations given as to how the wireless telegraph worked included the drawing of the shaded lines between a radio transmitter and receiver as if the cable existed in the air, but was invisible to the naked eye (Figure 4).

Thus, regardless of the functionality of broadcasting, the wireless network was visualized as a tool for communication between two points only. This point-to-point communication also served as a tool of imperialism; the radiotelegraph would transmit information from the colonisers to the colonized. One of the most influential examples of an imperial network is the Imperial Scheme of Marconi, which the Marconi Company proposed to build around the globe, in all British colonies.[7] The map of the Marconi chain, in fact, represents wireless telegraphy as an intersection of different lines. This map thus in no way resembles the radio broadcasting maps, which typically present stations as points with circles around them, thereby indicating their propensity to spread the information in all directions.

Second, even though wireless is open communication, in the air, accessible to everyone, initially it was imagined for private and secure communication. The popular music of the 1900s, for instance, pictured wireless telegraphy as a wireless telephone. Private correspondence over wireless featured prominently in Broadway songs that were issued for musicals and reached Europe as sheet music. These songs praised wireless and addressed the 'distant beloved,' one of the most common tropes in the popular song lyrics of the day.[8] The song "Love's Wireless Telephone," published in 1906, compared love to the wireless telephone, as it could transmit messages over long distances and 'seemed to be calling when we are alone.'[9] Not by chance, these early twentieth-century romantic fantasies of

---

6 Balbi, "Wireless's "Critical Flaw": The Marconi Company, Corporation Mentalities, and the Broadcasting Option."

7 Simon J. Potter and Jonathan Saha, "Global History, Imperial History and Connected Histories of Empire," *Journal of Colonialism and Colonial History* 16, no. 1 (2015): 1–14, https://doi.org/10.1353/cch.2015.0009; Heidi J.S. Tworek, "How Not to Build a World Wireless Network: German–British Rivalry and Visions of Global Communications in the Early Twentieth Century," *History and Technology* 32, no. 2 (2016): 178–200, https://doi.org/10.1080/07341512.2016.1217599; Dwayne R. Winseck and Robert M. Pike, "Communication and Empire. Media Markets, Power and Globalization, 1860–1910," *Global Media and Communication* 4, no. 1 (2008): 7–36.

8 Jon W. Finson, *The Voices That Are Gone: Themes in 19th-Century American Popular Song* (New York, Oxford: Oxford University Press, 1994).

9 Raymond W. Peck and Robert Hood Bowers, *Love's Wireless Telephone [Song]* (Detroit, New York: Jerome H. Remick & Co., 1906).

the transmission of feelings and emotions across long distances have today become reality with the advent of messaging devices that allow for the simulation and transmission of 'wireless smooches.'[10]

In this regard, the very fact that radio waves could spread information over large areas to many people was denied or appeared to be a major drawback. Marconi himself was one of the main figures to promote the understanding of wireless telegraph as a channel between the two points. He actively tried to prevent the phenomenon of broadcasting and insisted that a message could not be intercepted, which would cause him some embarrassment during one of his presentations of wireless. In 1903, the British magician Nevil Maskelyne publicly refuted Marconi's claims to have established secure and private communication. He managed in fact to broadcast a message of his own even as Marconi was demonstrating wireless telegraphy. Being sponsored by the Eastern telegraph company, hence supporting the interests of the competing electric telegraph, Maskelyne constructed a 50-m radio antenna with the specific aim of discrediting Marconi, who was transmitting from Cornwall to his great supporter Flaming at London's Royal Institution. The demonstration in question was supposed to show the capacity of Marconi's device to tune the transmission, however, during Flaming's actual lecture and before the presentation of the wireless telegraph, Maskelyne managed to broadcast his message. Anyone in the room who was conversant with Morse code immediately grasped what was happening. Maskelyne's message started with the word 'rats,' meaning empty boasting, nonsense, and went on to broadcast a limerick beginning 'There was a young fellow of Italy, who diddled the public quite prettily.'[11] This event was later described in the pages of the Times, as well as in various other newspapers, and was accorded a great deal of attention, with Flaming calling this savage attack 'scientific hooliganism.' In media history, Maskelyne gained the title of 'the first hacker.'

The initial development of wireless was thus governed by the idea of creating a secure point-to-point channel of communication, instead of opening it up to the public and engaging as many stations as possible. However, the technology was evolving very rapidly, even unpredictably, and this controversial development also found its reflection in the many names that wireless telegraphy acquired.

---

10 Paul Marks, "Kissing Device Lets You Send a Wireless Smooch," *New Scientist* 215, no. 2873 (2012): 20.

11 Raboy, *Marconi: The Man Who Networked the World.*

The controversial and contested names of wireless

Wireless or radio? Communication over radio waves developed in different modalities under different names: wireless telegraphy, radiotelephony, radio broadcasting, to name but a few. Media historians often tend to take for granted the terminology applied to the objects of study. However, as Rick Altman has argued, the terminology has major significance for our understanding of the technology because 'the innocuous process of naming is one of culture's most powerful forms of appropriation.'[12] In the early stages of wireless technology, the terms were not set in stone. Indeed, radio was perceived as an uncertain object, and its purpose and impact were not yet fully understood; therefore, it represented a new media that people 'did not yet know how to talk about.'[13] Some descriptions of radio transmissions even used the word 'wireless' in quotation marks, as if unconvinced of its literal meaning: without wires. For example, one did not receive a wireless telegram but a 'wireless' telegram instead.[14]

There were at least five words that were used interchangeably in various combinations in the 1910s and 1920s: telegraphy, wireless, radio, telephony, and broadcast. All of them often referred to the same technology, but different names brought to the fore distinct attributes of wireless communication and focused attention on some particular aspects while overlooking others. This uncertainty in linguistic usage also indicates the interpretative flexibility of the technology, which means that the technology was open to any and every interpretation and theoretically could have been developed in countless different ways.

Initially, the technology of conveying information using electromagnetic waves was considered an advance upon, and in part a replacement for, wired telegraphy. Therefore, the original term 'wireless telegraphy' explicitly referred to the telegraph and necessitated a discreet and private information channel. This attempt to keep everything under control, discreet, and point-to-point aligns with the critical issues of telecommunications history, as in electric telegraph or post.

The adjective 'wireless' referred to the method of transmitting information through electromagnetic waves, most frequently to the point-to-point transmission of messages.[15] In books and journals, wireless telegraphy was often depicted in the same way as the wired telegraph, with the dotted lines or small sparks of

---

12 Rick Altman, *Silent Film Sound* (New York: Columbia University Press, 2005), 16.

13 Benjamin Peters, "And Lead Us Not into Thinking the New Is New: A Bibliographic Case for New Media History," *New Media and Society* 11, no. 1/2 (2009): 13–30.

14 Mee, "The Unseen Telegraph Messenger Turning Electric Waves into Words."

15 Gabriele Balbi, "Telecommunications," in *Handbook of Communication History*, ed. Peter Simonson et al. (New York and London: Routledge, 2013), 209–22.

light somewhere in the air to indicate the unseen cables.[16] This connotation is noticeable in the adjective 'wireless' and its translations, such as *беспроводная* (rus.), *drahtlosen* (ger.), the adverb without wires and its analogues in Romance languages, such as *sans fils* (fr.), *senza fili* (it.), *san hilos* (sp.), and even acronyms like the French *TSF* (which stands for 'la telegraphie sans fils,' or telegraphy without wires). However, electromagnetic waves were soon discovered to be an open channel, and this negation of wires was no longer enough to contain all the meaning surrounding radio waves.[17]

The term 'radio' derives from the Latin word 'radius,' meaning the spoke of a wheel, a beam of light or a ray. Wolfgang Ernst drew attention to the fact that the word 'radio' primarily addressed specific properties of an electromagnetic field or the radial effect of the waves and not the manner in which the technology was used. The idea of sending out radio broadcasts was not embedded in the term from the beginning;[18] in fact, the terminology – wireless or radio – had already become a critical issue when the first international rules of this mode of communication were discussed by those attending the 1906 Berlin Conference. Radiotelegraphy, and not wireless, was chosen as an official international name for the technology in 1906.[19] However, even the translations of that same international convention were inconsistent with this decision. Yet the term "radiotelegraphy" best reflects the position and role of the technology in the 1910s; a capacity for open communication and radiation is acknowledged, but it is still seen as a form of telegraphy.

Another term that gradually emerged in the language of the 1910s was 'broadcast,' and it would prove to be one of the most enduring of this cluster of radio words. At first, it appeared as an adverb, but later became used as a verb and also as a noun. It assumed a gerundive form during the 1920s, when programs and transmissions became a regular activity and reference was made to 'broadcasting.' James F. Hamilton has traced the incidence of the word 'broadcasting' in the digital archives from its use in the late 1700s in agriculture and in evangelical publishing to the 1920s radio era and has discovered that the usage was closely

16 Mee, "The Unseen Telegraph Messenger Turning Electric Waves into Words."

17 Balbi, "Wireless's "Critical Flaw": The Marconi Company, Corporation Mentalities, and the Broadcasting Option."

18 Wolfgang Ernst, *Digital Memory and the Archive*, ed. Jussi Parikka (Minneapolis, London: University of Minnesota Press, 2013), https://doi.org/10.5860/choice.50-5970.

19 Le Département des Postes de l'Empire d'Allemagne, *Documents de La Conférence Radiotélégraphique Internationale (Berlin, 1906)* (Berlin: Le Département des Postes de l'Empire d'Allemagne, 1906).

linked with mystification.[20] This mystification implied the obscuring of human origins, decisions and actions: in other words, casting the seeds of the information as if they grew by themselves.

There were also terms that placed the emphasis upon the 'phone' traits of technology, with terms such as 'radiophone,' 'wireless telephony,' 'radiophony,' and others being deployed. The terms 'wireless telegraph' and 'wireless telephone' were often used interchangeably, though some sources brought out the technical differences between the technologies. The reference to a 'phone' was used mostly with two distinct meanings. First, with the development of communication using radio waves, experiments were conducted that focused on transmitting voice and speech, which were largely called 'radiotelephony.'[21] Second, the telephone incorporated both functions of sending and receiving messages at the same time, while the radiotelegraph required more energy for doing both. For instance, the battery was a fairly expensive element, and radio amateurs used only one for both the receiver and the transmitter; they had to switch to receive or send messages and were not able to perform both actions simultaneously. Therefore, the term 'wireless telephone' sometimes addressed the issue of point-to-point, plus point-to-point explicitly back again, meaning a two-way communication.

Figure 5 illustrates how the word 'radio' gradually suppressed all other terms during the period from 1910 to 1930. Both 'radio' and 'broadcasting' are evidently in the ascendant, while the lines for 'telegraph' and 'wireless' show a steep decline. This diagram presents only keywords that formed part of established and settled usage (e.g. 'wireless' and 'telegraphy' separately, rather than together as 'wireless telegraphy'), as these terms could feature in various different permutations ('wireless telegraphy,' 'radiotelegraphy,' etc.). This diagram does nonetheless capture important tendencies in the use of these terms, reflecting shifts in the general understanding of the medium itself.[22]

---

20 James F. Hamilton, "Excavating Concepts of Broadcasting," *Digital Journalism* 6, no. 9 (2018): 1–14, https://doi.org/10.1080/21670811.2018.1481762.

21 E.g., Arno Huth, *La Radiodiffusion. Puissance Mondiale* (Paris: Gallimard, 1937), 27–29.

22 No matter how instructive, this diagram has two important limitations that must be underlined. First, it only includes the English book corpus and therefore omits different variations of translations of these terms. Second, taken out of context, these terms could refer to other, non radio-related uses. For instance, the word 'telegraph' could refer to the electric telegraph. This fact complicates a rather simplistic explanation of the terminology change from this diagram, however, it still outlines important changes in the telecommunication and mass media that occurred in the 1910s–1930s regarding the use of the words 'radio' and 'broadcasting.'.

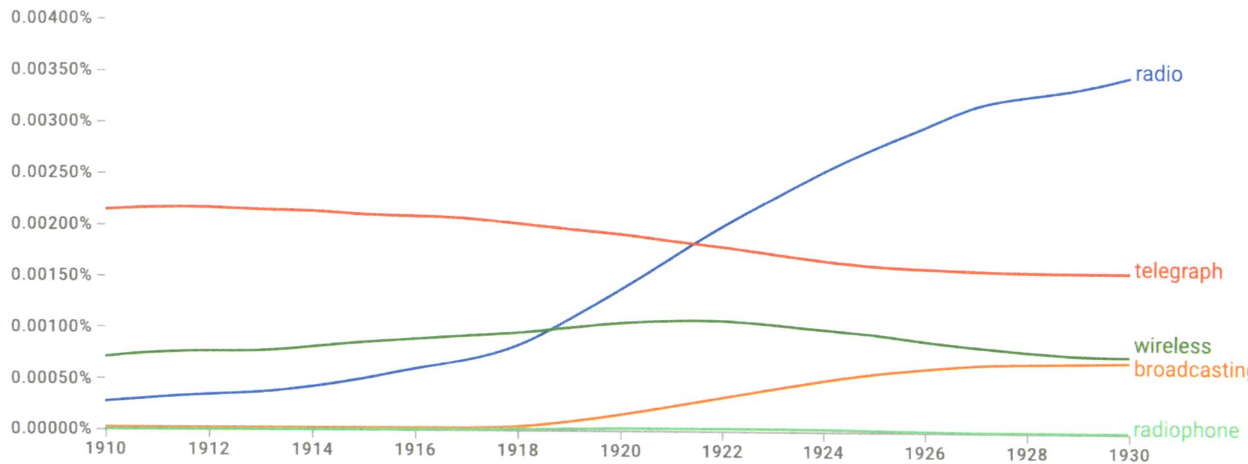

Figure 5: The frequency of the terms radio, telegraph, wireless, broadcasting, and radiophone, 1910–1930.[23]

23 Source: https://books.google.com/ngrams, (accessed April 26, 2024).

The problematic nature of these words and their various connotations reveal two crucial axes in nascent radio communication. The first is described in the literature as a point-to-point and one-to-many dichotomy,[24] meaning the intention either to keep communication secret or to spread information to the largest possible audience. Another axis for radio development concerns experiments with voice transmission versus Morse-code transmission. In other words, wireless or radiotelegraphy was always predominantly in Morse code and usually tended to be a point-to-point technology, with some exceptions involving one-to-many use. The technologies that allowed voice transmission were radiotelephony and radio broadcasting, and they were point-to-point and one-to-many, respectively (Table 1).

Table 1: Dominant uses of radio family terms in the 1910s.

	Morse coded	**Voice transmission**
Point-to-point	Wireless telegraphy	Radio telephony
One-to-many	Radio telegraphy	Radio broadcasting

The terminology regarding these new technologies regularly gave rise to confusion and misunderstanding. Apart from these keywords, there were also other references to radiotelegraphy; some examples are 'spark stations,' 'marconigram,'[25] and 'aerogram' business.[26] 'Cordless,' 'cable-free,' and other words address the absence of a physical connection. The list grows when translations into languages other than English and linguistic evolution are both taken into account. The official translation of the terms was not consistent either; for instance, in the various Hague conventions regarding the war, the word *radiotélégraphie* was translated from French as 'wireless telegraphy' rather than 'radiotelegraphy.'[27] Or another example: each year, the Marconi Company published the *Yearbook of Wireless Telegraph and Telephone* which did not even distinguish between these two media.

24 Gabriele Balbi and Juraj Kittler, "One-to-One and One-to-Many Dichotomy: Grand Theories, Periodization, and Historical Narratives in Communication Studies," *International Journal of Communication* 10 (2016): 1971–90.

25 "Monthly Miscellany," *Marconigraph* II, no. 20 (1912): 362.

26 "An Aerogram for Vienna," *The Wireless World* (September 1918): 340; "The Library Table," *The Wireless World* 4 (1919): 585–87.

27 A. Pearce Higgins, ed., *The Hague Peace Conferences and Other International Conferences Concerning the Laws and Usages of War: Texts and Conventions with Commentaries* (London: Cambridge University Press, 1909).

A perfect illustration of such a confusion is the following image of headphones. These were described as telephones, thereby adding another meaning to the term 'telephone'; it is evident, however, from the photograph that we would today call them 'headphones.'

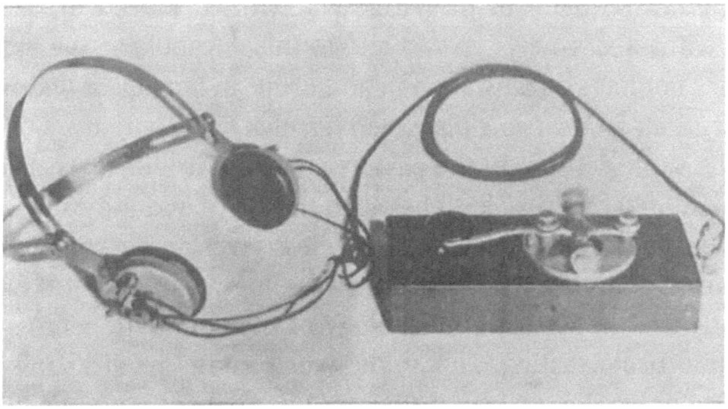

Figure 6: Headphones called 'telephone,' 1913.[28]

The description of the device only proves the point, as this was a part of a teaching device for radio operators. It uses the word 'telephone' to describe the headset: 'On the top cover of the box, a transmitting key is mounted, working with which the student sends an intermittent current corresponding to the signs of the Morse alphabet to the *telephone* (. . .) The head *telephone* serves to listen for transmitted signals.'[29]

Delegates at conferences, technicians, and ordinary users regularly criticized each other for their misuse of terms and, as a result, some of the principal engineers deliberately chose to avoid using any specific terms at all.[30] Journals also noted that wireless telegraphy had brought many words into the modern lexicon, and new coinages constantly appeared.[31]

Most of these terms have been (and still are being) re-invented and rethought in the late twentieth and early twenty-first century, resulting in the invention of new words (such as 'cordless telephones') and new applications under the same

28 "Teaching Device for Radiotelegraphists [Pribor Dlja Obuchenija Radiotelegrafistov]," *Herald of Wireless Telegraphy [Vestnik Telegrafii Bez Provodov]* (1913): 101.

29 Ibid., italics added.

30 Peter Pendleton Eckersley, "D. 40. N.8. Situation in Europe as Regards Wave-Lengths," in *Correspondance: Radiotélégraphie* (ITU Library & Archives, 1925).

31 "Language in the Making," *The Wireless World* (March 1917): 927–28.

names (such as 'wireless earphones'). Translations likewise proliferated, and the evolution of languages affected the terminology, disrupting its continuity. Thus, in Russian there are two ways of saying 'wireless' that are strikingly alike and yet still distinctly different: *беспроволочный* [besprovolochnyj] to refer to the 'wireless' of the early twentieth century as in wireless telegraphy and *беспроводной* [besprovodnoj] to address the present-day 'wireless' as in 'wireless networks.' Moreover, the contemporary epoch also witnessed a new term to add to this vocabulary: the adverb 'wirelessly,' which does not have a translation in all European languages and is sometimes expressed as 'without wires' (*sans fils*, *senza fili*, etc.), with only the German *drahtlos* reflecting the same sense. There is evidence of the use of the word 'wirelessly' in the 1910s, but it only really took off between the 2000s and the 2010s.

Reflecting on these different names and uses of wireless gives an overview of how complex and constantly evolving the wireless technology was and still is. Media historians frequently see the wireless technology of the twentieth century only as a precursor to radio broadcasting. Today, the word 'radio' itself usually refers to radio broadcasting only; along with television, it is a part of mass media history and is often referred to as a 'classic' example of media. Wireless, on the other hand, is seen as a type of connection (as in wireless lighting or wireless charging). The existence of different names also emphasizes different aspects of the newly developed technology, which also shows its 'transitional identity.'[32]

Thus, retrospectively, radio is defined through radio broadcasting only, however, initially it represented only one branch of these related technologies. Moreover, it was a quintessential component of the evolutionary development of radio technologies. Radio was not the only child of many parents but also an entire family with different children, and wireless is one of them.

This book uses both terms, 'wireless' and 'radiotelegraphy.' The term 'radiotelegraphy' is most apt in discussions of the transitional period between telegraph use and broadcasting of the 1910s and 1920s, which are at the heart of this book. It incorporates the contradictory ideas of radiating the message and telegraphing it point-to-point, which was precisely the stumbling block for the radio development of the 1910s. Moreover, the International Radiotelegraph Union, which is central to this research, also incorporated the word 'radiotelegraphy' in its title. From its choice of name we conclude that this was seen to be the most satisfactory way to address the subject of regulation and, consequently, the object of the current study as well. However, the choice was made also to sometimes use the

32 Simone Dotto, "Screening Radio or Broadcasting Cinema: How to Expose a Nameless Medium," in *A History of Cinema Without Names/3: New Research Paths and Methodological Glosses Udine* (Milano – Udine: Mimesis 299–316.

term 'wireless,' in the interests of maintaining consistency with the academic literature and thereby revealing a continuity between the wireless telegraphy of the 1900s and the wireless networks of today.

Patent wars and dividing the market

Very soon after the invention of wireless, the idea of wireless communication had transcended fantasy and theoretical treatises to become a commercially feasible and indeed valuable instrument of communication. Starting from the late 1890s, on the basis of theoretical research and practical inventions, entrepreneurial engineers began to set up their own companies in order to scale up the invention and seek profit.

The central figure in this wireless history is Guglielmo Marconi. By contrast with other inventors, Marconi was also an entrepreneur. Furthermore, he could call upon highly useful social connections within the British Parliament, thereby enabling him to sell his viable system of spark radio transmission to the British Post Office and to spread his invention throughout the world. The significance of Guglielmo Marconi in the history of communications can be compared to that of such contemporary figures as Steve Jobs or Bill Gates. He was charismatic, gave public speeches, managed his company ingeniously, and was able to reach a large part of the world market by conquering the communications market in the overseas territories of the British Empire.

There were, however, many rival firms. The initial period of wireless development from 1890 to the 1900s was characterized in general by a division of the global market. The most common instrument of that competition, which however caused more problems than clarity, was the fight and chase over patents. As wireless technology progressed at a rapid pace and simultaneously in many different countries, inventors patented very similar inventions in different national legislations, thus allowing them to compete on the international scene. Moreover, the process of getting a patent could itself take several years, in the course of which the invention of wireless technologies was debated. As regards the figure of Marconi himself, his invention was disputed by many. Marconi patented radio in 1896, however, before his patent had even become public, other inventors managed to submit their own patents. For instance, the syntonic circuit patented by Oliver Lodge in 1897 was crucial for establishing contact between the transmitter and the receiver, as it allowed the tuning of both to the same frequency.[33] Later,

33 Hugh G.J. Aitken, *Syntony and Spark. The Origins of Radio*, 1985.

in 1900, Marconi acquired his own tuning patent in response, however, the Lodge technique was more promising, and his earlier patent remained a bone of contention until Marconi managed to buy it in 1910.[34]

Another example is a patent battle between Tesla and Marconi that went on for decades. When both men had died, the US Supreme Court ruled that all of Marconi's radio patents were invalid and awarded the patents for radio to Tesla. However, it could not retrospectively rewrite the history– whoever was the 'right' inventor, the Marconi Company had still been a key actor in the development of wireless and radio at the beginning of the twentieth century. These patent wars over wireless and radio also proliferated because the technology was still evolving and changing: new techniques, apparatuses, and explanations appeared every day. And most importantly, the very notion of wireless had been evolving too.

The patent wars led to economic competition between different companies, known also as a 'commercial war' between them over the global market.[35] One of the most researched confrontations between engineering companies is the one between the German company Telefunken and the Marconi Company, which was of British-Italian origin. This confrontation frequently serves as a lens for researchers to study radio both on a global scale and also in a particular national arena.[36] Marconi's system of telegraphy was entirely based on spark transmission, which was the dominant system on the market during these decades. The primary approach was 'longer waves, taller antennas, higher power,'[37] which allowed radio waves to travel a longer distance and enabled ships to communicate. The greatest achievements of the Marconi company were associated with the great distance, such as, for example, crossing the Atlantic in 1902 (Marconi Archive, 1912).

Telefunken, however, could call upon the Slaby-Arco system of wireless telegraph, and upon the German electrical company A.E.G., which was highly protected by the German state. Because of the strong support from the military, the firm could both have a regular demand for its products as well as commercialize some technologies. As a result, Telefunken introduced a ground-breaking invention to the market, the quenched spark gap transmitter, which, in contrast to the unquenched gap, produced signals that were less noisy and much narrower in

---

34 Peter J. Hugill, *Global Communications since 1844: Geopolitics and Technology* (Baltimore, London: The Johns Hopkins University Press, 1999), 89.

35 William John Baker, *A History of the Marconi Company* (London, 1970), 96.

36 See Jesús Sánchez Miñana, *La Introducción a Las Radiocomunicaciones En España (1896–1914)* (Madrid: Fundación Rogelio Segovia para el Desarrollo de las Telecomunicaciones Ciudad Universitaria, 2004) on the history of radio in Spain.

37 Hugill, *Global Communications since 1844: Geopolitics and Technology*, 89.

bandwidth.[38] It also boosted the power that antennas could transmit, making it almost like a continuous wave, facilitating voice and music transmission.[39]

Despite their strong influence and advanced technology, the products of both companies had their pros and cons. Telefunken targeted land as space for communication and, in particular, communication with mobility, designing mobile radio sets for the army in the Herero War of 1904 and the Russo-Japanese War of 1904–1905. While tending to prevail on the battlefield, Telefunken was losing at sea, which became clear during the Russo-Japanese war, when the Russian Navy was supplied with the Telefunken devices, with the difficulties encountered in the use and maintenance of Telefunken wireless sets believed to be one of the reasons for the defeat of the Russians at the Battle of Tsushima (May 27–28, 1905).[40] After the war, the Russian state searched for a more reliable and effective technology and chose that of Marconi. In 1911, the Russian government thus negotiated an agreement with the Marconi Company, opening what was de facto a Marconi Company branch in Moscow to ensure a regular supply of the latest technological advances.[41]

The Marconi devices were, however, characterized by another peculiarity. They were distinguished by Marconi's idea of non-intercommunication, which meant that no other company could communicate with the Marconi devices. This decision was a business strategy adopted for a number of reasons: patent protection, economic, political, etc.[42] This policy of non-intercommunication has many similarities with the current media environment. Consider, for example, the incompatibility of software written for different operating systems, like Android or iOS. In the social landscape, this technical detail represented the attempt of Marconi to acquire a complete monopoly on the market, which often caused some disruptions in global communication networks and necessitated international regulations for radiotelegraphy.

---

38 Tapan K Sarkar et al., *History of Wireless* (New Jersey: Wiley-Interscience, 2006), 361.

39 Hugill, *Global Communications since 1844: Geopolitics and Technology*, 90.

40 Headrick, *The Invisible Weapon: Telecommunications and International Politics, 1851–1945*.

41 A.A. Glushhenko, *Place and Role of Radio Communication in the Modernization of Russia (1900–1917) [Mesto i Rol' Radiosvjazi v Modernizacii Rossii (1900–1917)]* (Saint-Petersburg: A.S. Popov Naval Radioelectronics Institute, 2005), 272–79.

42 Gabriele Balbi, "Marconi's Diktats. How Italian International Wireless Policy Was Shaped by a Private Company, 1903–1911", in *3rd Region-8 IEEE HISTory of Electro – Technology CONference: The Origins of Electrotechnologies, HISTELCON 2012 – Conference Proceedings* (2012), https://doi.org/10.1109/HISTELCON.2012.6487559; Michael Friedewald, "Telefunken vs. Marconi, or the Race for Wireless Telegraphy at Sea, 1896–1914," in *SSRN Scholarly Paper* (Rochester, NY: Social Science Research Network, 2012).

The patent wars, the many commercial innovations, and the capacity of radio waves to cross national borders easily and unwittingly created a novel socio-technical problem. Radiotelegraphy challenged existing systems and standards of national media infrastructure control, and called for political decisions to be taken regarding its organization and management. While the commercial scene was dominated by the British-based Marconi company, in the international political arena it was the German government that took the lead.

The formation of the International Radiotelegraph Union

In 1902, Prince Henry of Prussia travelled back from the US on the *S.S. Deutschland.* He attempted to send a radiotelegraphic message, but the operators of surrounding ships refused to forward it any further, obeying the aforementioned Marconi's non-intercommunication principle. The operators used Marconi devices, but the *S.S. Deutschland* was equipped with a Slaby-Arco apparatus.[43] The message was addressed, according to different sources, either to his brother, the Kaiser, or to US President Theodore Roosevelt[44] and represented a message of the utmost importance, with this incident having major significance for the overall development of radiotelegraphy; it intensified discussions about international regulations and is believed to be one reason behind the calling of the first international conference on radio.

In 1903, the German state invited delegates from ten countries to Berlin to discuss the most sensitive topics at the Preliminary Conference on Wireless Telegraphy. In addition to the US, representatives from nine European countries were in attendance: Germany, Austria, Spain, France, Great Britain, Hungary, Italy, Russia, and Belgium.[45] They were later also the most prominent representatives at the subsequent conferences, in 1906 in Berlin and 1912 in London, which thus established a Europe-centric focus for the ITU.

The capacity of radio to transcend national borders necessitated an emphasis on a shared conceptualization of the new technology. The 1903 conference reflected on this concept in the following way: 'Wireless telegraphy (. . .) projects its waves beyond the frontiers separating nations, so the protection necessary for its

43 Raboy, *Marconi: The Man Who Networked the World.*

44 ITU Archives, *What Is ITU?* (Geneva: Bureau International de l'Union télégraphique, 1974), 10.

45 ITU Archives, "Proceedings at the Conference," in *[Documents of the] Preliminary Conference on Wireless Telegraphy (Berlin, 1903). Translation of the Procès-Verbaux and Protocol Final by George R. Neilson* (London: George Tucker, 1903).

free development can only be secured with the concurrence of all the maritime nations by means of *an international understanding*.'[46]

Scholars interpret this aim to build an international understanding on radiotelegraphy as an ingenious way of presenting the more nuanced clash of British and German interests, which was also represented by the competition between the Marconi Company and Telefunken, respectively.[47] Germany found itself on an uneven playing field when it came to inventors, patents, and companies; despite its weak position, it attempted to battle the Marconi Company at the international conference. These conferences became an important geopolitical tool for Germans; as Orrin Dunlap asserted, 'the real purpose of the conference was to open the way for promotion of German wireless,'[48] and thus to challenge Marconi's non-intercommunication policy.

The importance of the competition between private companies in these international discussions was evident at the first conferences. For instance, in 1906, Great Britain and Russia made proposals (N 38 and N52, respectively) to the effect that governmental coastal stations would be allowed to charge ships a double price per radio telegram if the radio system employed on the ship differed from the one chosen by the country in question.[49] They were thus attempting to push the Marconi vision of radio into international regulations and were also supported by Italy, which acted as 'the international lawyer of Marconi Company.'[50] A detailed discussion ensued, with the German delegate, Mr. Sydow, underscoring that none of the other companies, such as de Forest or Telefunken, had demanded such a modification; the idea therefore aligned exclusively with Marconi's politics.[51] The proposal was rejected after a vote, thereby shaping the counter-Marconi policy evident in subsequent international regulations.[52]

This refusal to support the Marconi monopoly has frequently been interpreted as a victory for Germany over Britain in techno-diplomacy.[53] However, recent studies have shown that the Marconi Company's interests were not always

46 Italics added. Ibid., 6.

47 Hugill, *Global Communications since 1844: Geopolitics and Technology*; Headrick, *The Invisible Weapon: Telecommunications and International Politics, 1851–1945*.

48 Dunlap (1937), 148.

49 Le Département des Postes de l'Empire d'Allemagne, *Documents de La Conférence Radiotélégraphique Internationale (Berlin, 1906)*.

50 Balbi, "Marconi's Diktats. How Italian International Wireless Policy Was Shaped by a Private Company, 1903–1911."

51 Le Département des Postes de l'Empire d'Allemagne, *Documents de La Conférence Radiotélégraphique Internationale (Berlin, 1906)*.

52 Hugill, *Global Communications since 1844: Geopolitics and Technology*, 94.

53 For example, in Baker, *A History of the Marconi Company*.

synonymous with those of the British Government; this conference should therefore be seen only as harming Marconi's interests.[54] In fact, the 1906 conference made it plain that governments would have to take control of the development of wireless telegraphy instead of letting private companies and individual inventors determine the future market. As a result, the British state was forced to establish its own wireless service in 1908: the Post Office Wireless Telegraphy Section. In 1909, the Post Office purchased some strategic radio stations, which had previously been in the hands of the Marconi Company, such as the one at Crookhaven and those at the Lizard, Caister, Seaforth, and various others. In 1911, this section was upgraded to the Wireless Telegraph Establishment; later, during World War I, it was responsible for the British wireless service. Therefore, even though the Marconi Company still played an important role in the market and had a robust network, it did not have as exclusive a relationship with the state in the 1910s as had been the case in the 1900s.[55] These conferences illustrate that the development of radiotelegraphy became a politically crucial topic and could no longer depend only on the competition between commercial companies. Indeed, the commercial companies did not even have a vote at the international meetings. As a result, even Marconi, a famous inventor of radiotelegraphy with a series of patents to his name, had to follow the top-down policy that resulted from the adoption of international regulations by the various countries.

After the 1903 preliminary meeting and the 1906 conference, an international arena for negotiations on the future of radiotelegraphy was formed, with the Radiotelegraph Union created in 1906 as one of the outcomes of the conference. The Radiotelegraph Union was not an independent international organization; the term referred to the group of countries that had adhered to the various Radiotelegraph Conventions.[56] The name aligned with other unions existing at the time, such as Postal Union or Telegraph Union, and the word Union in fact referred to a collective effort to unite the communication networks.

The 1906 conference suggested associating this arena with the already existing Telegraph Union. The ITU is currently located in Geneva, Switzerland, and it is an agency of the United Nations; at the beginning of the twentieth century, it was based in Bern, Switzerland. This Union was created in 1865 to regulate and unite the national networks of electric telegraph so that information could easily

---

54 Bruton, "Beyond Marconi: The Roles of the Admiralty, the Post Office, and the Institution of Electrical Engineers in the Invention and Development of Wireless Communication up to 1908 (PhD Thesis)."

55 BT Archives, "Records of the Wireless Telegraphy Establishment; 1910–2001. Administrative History" (1910).

56 Lommers, *Europe – on Air: Interwar Projects for Radio Broadcasting*, 57.

cross borders[57] and was the first international organization in the world, established even before the Red Cross; it was and still remains one of the most important actors in the telecommunications industry. The alignment of radiotelegraphy with the Telegraph Union was therefore not in itself surprising. From its establishment in 1865, the Telegraph Union had indeed been the most important arena in which to discuss transnational communication. The national states were able to apply a new means of communication more effectively by allowing it to transcend national borders and extend its correspondence with neighboring countries according to international agreements. With the invention of radiotelegraphy, the questions of intercommunication and the allocation of spectrum and transnational flows had already been discussed at ITU conferences beginning in the 1890s in regard to the electric telegraph. The delegates to the first radio conferences had also been present at electric telegraph conferences; for instance, J.J. Perk, the minister of the colonies of the Netherlands and delegate from the Dutch East Indies, was present at the conferences of Paris (1890), Budapest (1896), London (1903), Lisbon (1908), and also at other non-radio conferences.[58] Furthermore, some regulations of the electric telegraph also concerned wireless communication and, as a result, the conventions on the wired telegraph were sometimes falsely considered radiotelegraphic conventions.[59] Thus, there were already many connections with the ITU before the radiotelegraph conferences had even taken place.

The main administrative body of the ITU was and still is the Bureau. Its principal activity consists of organizing international conferences, where the representatives of state-members discuss modifications to international regulations; as a result of these negotiations, the international convention is signed. After the Berlin conference in 1906, it was decided to divide the Bureau into two independent sections: telegraph and radiotelegraph, a task that took some time and effort.[60] The radiotelegraph component was to be in charge of the gathering and dissemination of information among state-members of the Radiotelegraph Union in the interval between the conferences. The International Radiotelegraph Union (IRU) was therefore never actually established as an organization but instead

57 G. Fari, S. Balbi, and G. Richeri, "A Common Technical Culture of Telegraphy: The Telegraph Union and the Significance of Technological Standardization, 1865–1875," in *HISTory of ELectro-Technology CONference, Third IEEE* (2012).

58 ITU Archives, *Documents de La Conférence Radiotélégraphique Internationale (Londres, 1912)* (Bern: Bureau International de l'Union télégraphique, 1913), 102.

59 Bureau des longitudes, "Electro-Optique (Suite)," in *Annuaire Pour l'an 1916 Publié Par Le Bureau Des Longitudes* (Paris: Gauthier-Villars et Cie, 1916), 454.

60 ITU Archives, *L'Union Télégraphique Internationale (1865–1915)* (Bern: Bureau international de l'Union télégraphique, 1915), 38.

formed its own administrative unit as a part of the ITU Bureau. Even if some authors refer to these sections as two different organizations (i.e. the International Telegraph Union and the International Radiotelegraph Union[61]), in fact, radio and telegraph administrations shared the infrastructure, materials, and resources in the ITU Bureau in Bern. In 1932, at the Madrid conference, they officially merged, which was a logical and even predictable scenario. As early as 1906, when discussing the future of the radiotelegraph office, the French delegate noted that the telegraph and radiotelegraph unions would undoubtedly merge in due course.[62]

Apart from the responsibility to organize the conference, the ITU Bureau was also charged with collecting, coordinating, and publishing information on radiotelegraphy, according to Article 13 of the 1906 convention.[63] All the maintenance costs of the Bureau fell on the contracting countries, at a maximum of 40,000 francs a year,[64], with the Bureau thus responsible for facilitating international collaboration. It is important, however, to highlight that the ITU Bureau was not in charge of implementation of the international conventions; the Conventions delegated the responsibility for implementing international regulations to the administrations of the various states adhering to the convention. The ITU Bureau therefore only organized the conferences, gathered data from state administrations and commercial companies, and published overviews on the general situation in radiotelegraphy. At the same time, each administration collected material from the ITU and disseminated it across its networks, discussed international matters, proposed amendments, adopted national legislation, and controlled the use of radiotelegraphy in its own country. It should also be stressed that the ITU influence expanded beyond the list of regular subscribers, as most of them passed the ITU information to their own networks. For instance, on March 26, 1914, the Board of Trade in the UK officially requested that copies of supplements published by the ITU Bureau be regularly sent, a request which was later approved and included in the distribution list; the General Post Office was, in fact, therefore functioning as a mediator of the messages coming from the international agenda.[65] In fact, the documents produced by the ITU were spread extensively among the countries' members and consequently

61 E.g., Victor Meyer, *L' Union Internationale Des Télécommunications et Son Bureau* (Bern, 1946).

62 Le Département des Postes de l'Empire d'Allemagne, *Documents de La Conférence Radiotélégraphique Internationale (Berlin, 1906).*

63 ITU Archives, "Convention Radiotélégraphique Internationale (1906)," in *Documents de La Conférence Radiotélégraphique Internationale (Berlin, 1906)* (1906), 347.

64 Le Département des Postes de l'Empire d'Allemagne, *Documents de La Conférence Radiotélégraphique Internationale (Berlin, 1906).*

65 BT Archives, "POST 30/2576B. Transmission by Wireless Telegraphy of Time Signals and Weather Reports (Change to 1913–1914)" (London, 1913).

ended up in different publications, reviews, and national archives. There was not only exchange of information but also constant reproduction and re-appropriation. The ITU was exploiting the so-called knowledge infrastructure, defined as a 'robust networks of people, artifacts, and institutions which generate, share, and maintain specific knowledge about the human and natural worlds.'[66]

Uses and users of radiotelegraphy

German playwright Bertolt Brecht, reflecting on the initial period of radio, remarked that 'it was suddenly possible to say everything to everybody but, thinking about it, there was nothing to say.'[67] Raymond Williams also noted that transmission and reception 'as abstract processes' preceded the content creation for both radio and television. He wrote, 'it is not only that the supply of broadcasting facilities preceded the demand; it is that the means of communication preceded their content.'[68] This was altogether true of radiotelegraphy; not only was the technology evolving, but so too was the content of the messages and its uses.

At first, the advantages of communication without wires seemed to show themselves to best effect at sea, where ships in motion had no other ways to communicate with land and with each other. Radiotelegraphy offered two highly important features: the ability to communicate while moving and no requirement for laying cable. In most countries, the number of radiotelegraph stations on board exceeded the number of shore stations by 1912, according to the ITU statistics on radiotelegraph stations.

In 1912, the US and the UK had the greatest number of radiotelegraphic stations: 873 and 845, respectively. The number of radiotelegraphic stations in other countries was considerably lower; Germany (including all its protectorates) had 399 stations, France was operating with 246, while Italy and the Russian Empire followed with 142 and 104 stations, respectively. Canada (97), Brazil (89), the Netherlands (80) and Austria (67) were also in the top ten leading countries according to the number of stations, while all other countries had fewer than 50 stations each (see Appendix A). Figure 7 illustrates the total number of open radiotelegraphic stations in these ten leading countries. Different colors indicate the num-

66 S. Edwards, P.N. Jackson, S. Chalmers, M, Bowker, G.C. Borgman, C.L. Ribes, and D. Calvert, *Knowledge Infrastructures: Intellectual Frameworks and Research Challenges* (Ann Arbor, MI: University of Michigan School of Information, 2013).

67 Brecht, "Radio as a Means of Communication. A Talk on the Function of Radio," 24.

68 Raymond Williams, *Television: Technology and Cultural Form* (New York: Schocken Books, 1975), 25.

ber of stations erected on coasts and ships, respectively. Most of these nations emphasized ship-to-shore radio communication in contrast to that between land stations.

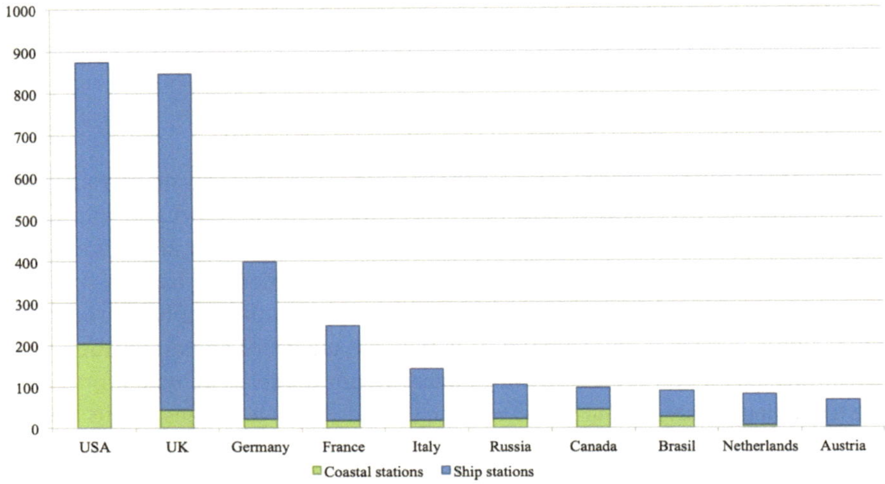

Figure 7: The total number of radiotelegraphic stations, 1912.[69]

Ship stations accounted for an overwhelming proportion of the total number of radiotelegraphic stations. On average, about 80 per cent of the total number were erected on ships in 1912. Only countries covering vast land masses (such as Russia, the US, Brazil and Canada) developed more coastal stations. The European countries used a limited number of terrestrial stations, as their territories were reasonably small and operators could easily cover the major part of their coastlines.

The proliferation of radio in sea transportation had also affected how radio was conceptualized. While radio altered perceptions of the sea, rendering it a more accessible and researched space, the sea had in its turn shaped the understanding of radio. In particular, radio was seen as an international technology because it not only transcended national borders but also was inherently a part of the sea, which had been defined as an international space (the so-called *mare liberum*) in a seventeenth-century agreement. Radio was believed to be treated

69 Based on ITU Archives, *Statistique Générale de La Radiotélégraphie Dressée d'après Des Documents Official. Année 1912* (Bern: Bureau International de l'Union télégraphique, 1914).

equally as an internationally shared space and place of communication.[70] Therefore, radio was seen as a common good among maritime nations.

Only those countries that had access to the sea were invited to the first radio regulation conferences; countries that did not participate in maritime matters did not even have a say in the future of radio. The only exception, one which does however prove the rule, was the case of landlocked Switzerland. In 1906, British, French, and Bulgarian delegates proposed to entrust the ITU in Bern with the task of handling all administrative work on the international legal issues of radio communication between the conferences. The proposal was met with confusion and consternation, as Switzerland not only did not have sea access but also had not even been invited to the conference. However, the supporters of this proposal assured everyone that there would be no objections from Switzerland, and it would take all responsibility, as it already managed the entire infrastructure of the ITU Bureau.[71] The fact that Switzerland did not have sea access theoretically also put it in a more neutral position concerning radio.

Aside from military applications and helping with maritime navigation, wireless has generally been seen as a telegraphic business. The radiotelegraph was used mainly for publishing newspapers on board and for private correspondence. The primary interest of the wireless companies in having radio apparatuses on passenger liners was purely commercial, and the passengers were treated as the customers of the service. Radio operators frequently exchanged messages with shore stations in batches at night, and the coastal stations would forward it further inland, either by wireless or more often by cable. These private messages were often addressed to loved ones and contained information of a personal nature.

The 1910 song 'Send Me a Kiss by Wireless' is a perfect example of this commercial application of wireless. It presented radiotelegraphy as a communication means for private correspondence on board a ship. The cover captured a cherub in the clouds chasing the vessel at sea, which reflects this common image of both love and wireless as something coming through the air (Figure 8). The radio message was visualized as a physical letter in the cherub's hand that emitted sparks and lightning, which corresponds to the electrical nature of the radiotelegraph and spark transmitters. The cherub was pictured as a postman, which therefore refers to the idea of wireless telegraphy as a telecommunication tool, secure and point-to-point. The cover art of this song manages to address the most common

---

70 P. Fauchille, "Le Domaine Aérien et Le Régime Juridique Des Aérostats," *Revue Générale de Droit International Public* 8 (1901): 414–85.

71 Le Département des Postes de l'Empire d'Allemagne, *Documents de La Conférence Radiotélégraphique Internationale (Berlin, 1906).*

Figure 8: The cover of the song 'Send Me a Kiss by Wireless,' 1911.[72]

issues present in the imaginary of radiotelegraphy: its use in the sea, communication across the air, and private correspondence.

Private communication over wireless was also often visualized as a luxurious part of travelling. For instance, a drawing from 1912 depicted a servant bringing a message to a gentleman who was relaxing on his deck-chair under a blanket with a book (Figure 9). The scene represented the ship's deck, and the message was

72 Earl Caroll, *Send Me a Kiss by Wireless [Song]* (New York: Jos. W. Stern & Co, 1911).

served on a plate. The title of the drawing was 'Delivering a wireless message,' even though there was nothing in the picture that indicated its wireless nature; instead, it was very physical and tactile for the final customer. The wireless content was a message written on the piece of paper, and the wireless infrastructure was completely hidden. In this instance, the role of radiotelegraphy was purely seen as a tool of entertainment.

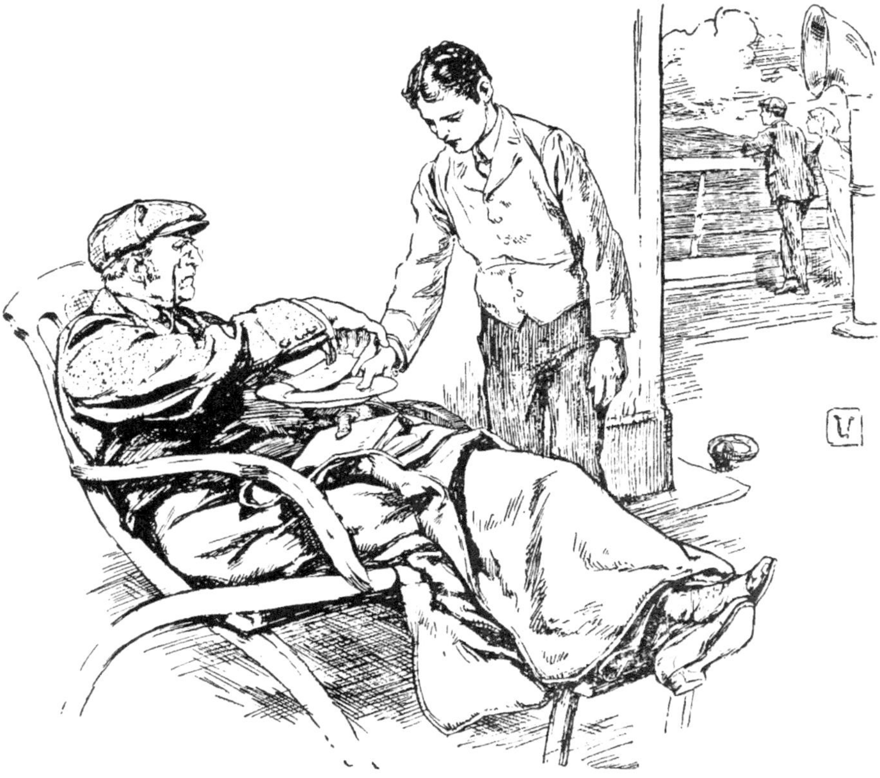

Delivering a wireless message

Figure 9: Delivering a wireless message on the ship, 1912.[73]

73 Francis A. Collins, "Delivering a Wireless Message," in *The Wireless Man: His Work and Adventures on Land and Sea* (New York: The Century Co, 1912).

The technological aspect of radio also provoked mysterious explanations. Radio waves were invisible, and the auditory signal was barely comprehensible. It did not yet involve the straightforward transmission of voice; messages arrived as auditory signals in a series of strange beams and sometimes were even faulty – on account of interference. Only a skilled radio user could easily decode Morse Code into a message. In his 1902 novella *Wireless*, Rudyard Kipling marvellously captured this inexplicability and offered a supernatural perception of radio devices:

> Have you ever seen a spiritualistic seance? It reminds me of that sometimes––odds and ends of messages coming out of nowhere––a word here and there––no good at all.[74]

Radiotelegraphy coexisted with various fantastical speculations, predictions, and dreams that were also correlated with various new, spiritual, and inexplicable technologies.[75] Still, however inexplicable the use of radio was, the experience of playing with the devices and messages was in a way seductive. Gregory Whitehead explains this attraction in the following manner:

> the radio signal as intimate but untouchable, sensually charged but technically remote, reaching deep inside but from way out there, seductive in its invitation but possibly lethal in its effects.[76]

Not surprisingly, therefore, the early days witnessed a proliferation of experiments with different content. For instance, one of them was a ship-to-shore radio transmission of electrocardiograms and X-ray images to advance patient care. The accuracy of this biological data was crucial, and early radio technology could not provide a sufficient degree of precision. Man-produced interferences and atmospheric conditions led to distortion in the ECG configuration that could endanger the patient's life; this use of radio was therefore concluded as unsuccessful.

---

74 Rudyard Kipling, "Wireless (1902)," in *Traffics and Discoveries* (Cornwall: House of Stratus, 2008), 164–85.

75 Simone Natale, "A Cosmology of Invisible Fluids: Wireless, X-Rays, and Psychical Research around 1900," *Canadian Journal of Communication* 36, no. 2 (2011): 263–75; Simone Natale, *Supernatural Entertainments: Victorian Spiritualism and the Rise of Modern Media Culture* (University Park, Pa.: Pennsylvania State University Press, 2016); Simone Natale and Gabriele Balbi, "Media and the Imaginary in History: The Role of the Fantastic in Different Stages of Media Change," *Media History* 20, no. 2 (3 April 2014): 203–18.

76 Gregory Whitehead, "Out of the Dark. Notes on the Nobodies of Radio Art," in *Wireless Imagination: Sound, Radio and the Avant-Garde*, ed. Douglas Kahn and Gregory Whitehead (Cambridge (MA), London: The MIT Press, 1992), 253–65.

Conclusions

The 1900s were rife with new applications of wireless. Yet it was a mysterious and unknown technology, with much still to be explored. There were many contradictions in the level of understanding so far achieved of wireless technology, and also many controversial names that were used interchangeably: wireless, radio, telegraphy, telephony, broadcasting. The technology was evolving month by month. However, despite its ability to radiate messages, the technology was mostly perceived as a substitute for the wired cable, and therefore considered a point-to-point communication tool. Any capacity to broadcast the signal was seen as a great disadvantage of the communication, and the main effort expended in the developing of the networks was devoted to organizing the tuning of the devices in such a way as to keep communication private and one-to-one only.

The early 1900s witnessed a boom in the wireless industry: new companies kept opening, patenting their inventions, and fighting over different markets. It was a commercial epoch of wireless: the main driver was profit. Companies were mostly targeting a wealthy public, and the radiotelegraph services for the most part targeted private correspondence.

The networks of radiotelegraphy also sought to expand over the globe, and the companies therefore always worked under the patronage of different political powers. In Europe, the fiercest competition developed between the UK and Germany, the Marconi Company and Telefunken respectively. The networks of these companies extended across the globe, in an obviously imperialist perspective – that is to say, covering in the main the territories of their respective empires. Despite being global, the networks were hardly interconnected and did not facilitate communication between rivals.

This competition resulted in an international struggle, which led to the formation of the International Radiotelegraph Union, which aimed to solve transnational issues through the management and control of wireless communication. Various European powers signed a convention that would regulate communication between different networks, and thus facilitate global interconnectedness. The countries in charge were those that had access to the sea, as this space stood most in need of mobile communication without cables. The sea was also seen as an international space that was to be regulated through a shared effort on the part of the different powers, and management of wireless telegraphy consequently inherited the same attitude. Wireless was already seen as a powerful commercial instrument for private communication, whereas its potential for global networking and contributing to the international public good was yet to be revealed.

Chapter 2
Wireless for a global audience, the early 1910s

This chapter depicts an international era of radiotelegraphy in the early 1910s, which was very bright and, unfortunately, also very short. Yet it did much to shape the international applications of wireless and anticipated many globalization patterns evident in the wireless communication of today.

First, this chapter highlights the importance of the 1912 *Titanic* tragedy for subsequent developments in radio. It analyzes the ITU 1912 Conference materials and demonstrates how the international regulations on wireless were transformed, determining the subsequent transnational understanding of radio communication. It further explores the political economy of radio communication through a close examination of the transnational organization of the market and of global companies. It then goes on to present important transnational cases and projects using radiotelegraphy, such as international time signals and weather reports. Finally, it explains the part played by radio amateurs in introducing wireless telegraphy to the wider society.

The Titanic reveals the flaws of the wireless industry

The distress signal interrupted the silence on the cold expanses of the Atlantic ocean. 'This is the *Titanic*! This is *Titanic*!' The telegraph operator Jack Philips sent the wireless message over and over again calling for rescue. Heroically he remained on board until the ship went down, tirelessly sending the very well-known call for rescue at that time, CQD, and the recently introduced SOS. The vessels around the *Titanic* received these signals and rushed to the rescue, exchanging hundreds of messages among each other, clarifying details and sharing the *Titanic*'s location. The signals from the *Titanic* were very faint, until they disappeared completely when the ship was drawn down into the ocean. Only one vessel made it in time to pick up some survivors.

The *Titanic* was branded as the greatest ship ever. The equipment on board the *Titanic* was described as the most advanced and powerful of its time; it used the wireless apparatuses of the Marconi Company. The photographs of the *Titanic* itself featured a powerful image of the wireless antenna as a clearly visible mast with four parallel wires (see Figure 10). The radio operators had also been inten-

Open Access. © 2024 the author(s), published by De Gruyter. BY-NC-ND This work is licensed under the Creative Commons Attribution-NonCommercial-NoDerivatives 4.0 International License.
https://doi.org/10.1515/9783111202327-003

sively trained to receive messages about meteorological conditions, ice reports, and the presence of fog or derelicts.[1]

Figure 10: The *SS Titanic* with the radiotelegraph aerial array, 1912.[2]

However, despite the advanced equipment and staff training, the wireless could not help to save all the passengers and crew. How did that happen?

On the night of April 15, 1912, the *Titanic* struck an iceberg. Around 10.15 pm, Captain Smith entered the wireless room and instructed the operators to call for assistance. The *Titanic* had two radio officers on board: 25-year old John (Jack) Philips from Surrey, who had sailed as radio officer since 1906 and previously worked as a senior operator on a high-power wireless station in Ireland, and 22-year old junior operator Harold Bride from Nunhead, who had a working experience of about a year. Both had undergone a special course of training to be qualified operators and were experienced professionals.[3]

Most probably, the radio operator on duty, Jack Philips, did not receive any warnings about the proximity of icebergs because he was concentrating on com-

1 Michael Hughes and Katherine Bosworth, *Titanic Calling: Wireless Communications during the Great Disaster*, ed. Michael Hughes and Katherine Bosworth (Oxford: The Bodleian Library, 2012), 23–24.

2 Marconi Calling Digital Archive, "SS Titanic (1)" (Chelmsford, 1912), https://wayback.archive-it.org/org-467/20121016163838/http://www.marconicalling.com/museum/html/objects/photographs/large_image/large_image-type_d__t04582.html.

3 Hughes and Bosworth, *Titanic Calling: Wireless Communications during the Great Disaster*, 24.

munication with shore stations and did not want to be distracted by other messages. Just before the disaster happened, he transmitted to other operators: 'Shut up, shut up. I am working Cape Race' (a shore station in Newfoundland), which resulted in his missing information about icebergs.[4]

When the *Titanic* struck the iceberg, the radio operators were instructed immediately to send out distress signals, which were soon picked up by ships in proximity and then relayed rapidly to other ships. The operators used the emergency call sign for Marconi operators (the CQD) and also a recently introduced international distress call, SOS. All operators should have readily understood both; CQD was a well-known code and SOS had been used sporadically for about three years. Indeed, it had been the official distress call since 1908 (under the terms agreed at the 1906 Berlin Convention on Radiotelegraphy). Junior operator Harold Bride, who survived the disaster, wrote a statement just after he arrived in New York, where he recalled joking with Jack Philips about the distress calls: 'Send SOS' I said, 'It's the new call, and it may be your last chance to send it.'[5] Sadly, it was indeed Philip's last sending of a distress call.

Many vessels plainly did receive the call. The tragedy also took place in very close proximity to a large vessel, the *Californian*, which could have saved some *Titanic* passengers, but its radio operator was off duty and therefore did not hear the distress signal. Only one ship reached the site of the sinking in time to pick up survivors: the *Carpathia*.[6]

Figure 11 represents how radiotelegraphy was thought to have worked on the *Titanic*. An image published in the Chicago-based newspaper the *Day Book* explained that the *Titanic* broadcasted the SOS signal to get help, with the lines drawn as long bolts of lightning, which indicated electric communication. As Thibault noted, along with the wave, the bolt was one of the most common ways to visualize the radiotelegraph.[7] What is peculiar about this image is that broadcasting is here envisioned as numerous point-to-point channels between the *Titanic* and the other, surrounding ships, as if the message could have reached only them. This also could be interpreted as a transitional period in understanding ra-

4 ""Titanic" Disaster. Hearings before a Subcommittee of the Committee on Commerce, 62 Congress, 2 Session, Pursuant to S. Res. 283, Directing the Committee on Commerce to Investigate the Causes Leading to the Wreck of the White Star Liner "Titanic"." (United States Senate, 1912), 735–40.
5 As cited by Hughes and Bosworth, *Titanic Calling: Wireless Communications during the Great Disaster*, 115.
6 Hughes and Bosworth, *Titanic Calling: Wireless Communications during the Great Disaster*.
7 Ghislain Thibault, "Bolts and Waves: Representing Radio Signals," *Early Popular Visual Culture* 16, no. 1 (24 May 2018): 39–56, https://doi.org/10.1080/17460654.2018.1472621.

diotelegraphy: both the notion of point-to-point and that of one-to-many are present.

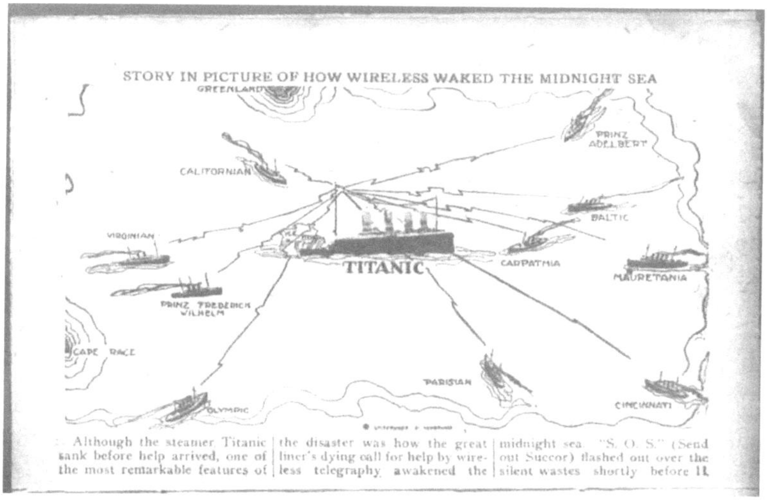

Figure 11: The *Titanic* broadcasting an SOS-signal point-point to other surrounding ships, 1912.[8]

When Jack Phillips desperately kept tapping 'dot, dot, dot, dash, dash, dash . . .', he understood only too well that wireless was the only key to salvation. Unfortunately, in the world at that moment, the potential of wireless communication for saving hundreds and thousands of people was not yet fully realized. Although most ships had wireless devices on board, the staff and companies treated this technology very casually. Wireless technologies were for sending jokes, letters, and news, thus as more of a source of entertainment and fun. They were not perceived as a channel for anything serious, because the radio waves were unencrypted communication channels and open to anyone. Apart from the commercial successes of the different companies in entertaining a wealthy public, and the military developments undertaken by the Navy, the communication space was also full of the fun and joy of experimenting with the technology. In their spare time professional operators on ships would even play games 'on air.'

8 "Story in Picture of How Wireless Waked the Midnight Sea," *The Day Book. (Chicago, Ill.)*, April 17, 1912.

The case of the *Titanic* conveyed a message to the whole world: lawlessness in wireless spaces should be stopped. It had revealed how the existing technologies could be used to solve old problems, if used correctly at the right moment. Wireless technologies could, it was clear, unite nations and continents. The sheer scale of the losses on board the *Titanic* shocked the entire world. Indeed, 1912 represented a watershed for radio development, transforming the international debate on radiotelegraphy and having a major impact on international regulations and on the general approach to this new means of communication.

An investigation into the sinking of the *Titanic* was organized swiftly, the day after the *Carpathia* had docked in New York. The investigation was based on the radiotelegraph logbooks kept by wireless operators as a record of all messages sent and received: the so-called procès-verbaux. Several radio operators of surrounding ships also provided evidence. The examination revealed that radio communication could have been used more wisely and could have allowed more passengers to be rescued, if only it had not been considered purely as an entertainment tool. These details, and the story of *Titanic* in general, proved the major importance of radio in the open sea for rescue and had a huge impact upon subsequent political events and projects.

The International Radiotelegraph Conference in London

Shortly after the *Titanic* disaster, the International Radiotelegraph Conference took place in London from June 4 to July 5, 1912. It was the third international meeting to address the topic of radio, but the first to be organized by the ITU and not by the German government. It held nine plenary sessions, ten meetings of the Rules Commission, six meetings of the Tariff Commission, six meetings of the Commission Drafting Committee, and several sessions of various sub-commissions (ITU Archives, 1912d). Marconi himself also "entertained" the delegates at the opening of the Marconi radio equipment factory in Chelmsford (Marconi Archives, 1912).[9]

The 1912 ITU conference was a landmark for radio development. It greatly expanded the list of the international rules for the use of wireless telegraphy, which lasted for many decades. Delegates, previously hesitant to impose any restrictions whatsoever on radiotelegraphy, were inspired to resolve the complications attendant upon wireless communication because of the recent *Titanic* tragedy. Their de-

9 ITU Archives, "Rapport de Gestion de l'Union Télégraphique (Section Radiotélégraphique)," 1912.

cisions affected the general understanding of radiotelegraphy and imposed a particular vision of it for future development.

The 1912 conference had been scheduled long in advance. As with any international conference, the Bureau collected issues for the agenda for many years. The representatives of the various countries sent in their packages of documents that included suggestions for international regulation, so that their counterparts could get familiarized with them in advance, however, many of those questions were rethought or postponed due to the more urgent topics of discussion emerging after the *Titanic* disaster. Along with the *Titanic* investigation and other international conferences happening in 1912, the delegates of the ITU conference attempted to learn a lesson from the *Titanic* tragedy. The President of the conference called for a revision of regulations in light of the recent event:

> The recent '*Titanic*' disaster has drawn everyone's attention to the use of wireless telegraphy in maritime disasters, and one of the most important tasks of the present Conference is to consider whether the possibility of providing rescue can be improved by amending the provisions of the Convention and the Regulations.[10]

Thus, delegates put all their efforts into designing regulations in a manner that could turn radiotelegraphy into a sustainable, accessible, and useful communication tool, one that was readily and swiftly deployed for rescue. That meant that the *Titanic* brought a completely new vision for radiotelegraphy on to the international agenda: it demonstrated how important it is not just to send a message, but to be heard. The *Titanic*'s case underlined the importance of collaboration, a collaboration that had to be transnational and supported by different actors in the wireless industry. Moreover, it also revealed the importance of clear organization of the radio spectrum and the harmful or even lethal impact of interferences on air.

There were many outcomes of the 1912 conference. The regulations doubled in size, expanding from an 8-page document to 16 whole pages.[11] They tried to cover every possible aspect of wireless telegraphy, and three of those outcomes were particularly interesting and influential for the future development of wireless.

First of all, radiotelegraphy was transformed from a means of entertainment to a component of the safety infrastructure. According to the convention signed at this conference, ships were required to have a radio set and operators on duty. This rule shows how drastically the understanding of radio had changed. Radio was no longer considered an instrument of commercial entertainment but was

10 ITU Archives, *Documents de La Conférence Radiotélégraphique Internationale (Londres, 1912)*, 255.

11 ITU Archives, "Convention Radiotélégraphique Internationale (1906)"; ITU Archives, *Convention Radiotélégraphique Internationale (Londres, 1912)* (1912).

supposed to safeguard a ship and its passengers by enabling the crew to remain in constant communication with stations on land and also with other vessels. Moreover, radio ceased to be an exclusive and luxurious appurtenance of travel. All vessels would henceforth be obliged to install wireless stations on board and hire an operator with a first-class radiotelegraph certificate. Furthermore, as ship-to-shore communication had to be facilitated, the regulations insisted on placing fewer restrictions upon the technology. This meant that the imperative of safety allowed Marconi's monopoly to be entirely broken, as all ships were now legally obliged to be able to communicate with each other.

Second, the regulations asserted tighter control over the radio spectrum and likewise its clear organization. The discussion of the SOS-signals made this particularly evident. The signal had been in use since 1908, however, most ships used another one, which was diffused among the radiotelegraphic stations of the Marconi Company. It was 'CQD,' which stood for two acronyms: CQ, 'to all stations,' and D, 'distress.' This message was quite difficult to remember as it was a complicated combination of dots and dashes (-.-. --.- -..). The newly adopted international distress call, the SOS signal, was symmetrical in Morse code (... --- ...) and shorter, which therefore made it easier both to remember and to send. In fact, the *Titanic* case is also notable from the fact that the liner was sending both signals at once: the operator initially sent out the CQD-code, but the second radiotelegraph operator suggested that they try a recently introduced international distress call, SOS, almost as a joke. This underscores once again that, despite being already an official distress call, the SOS-signal was not very widely used. Thus, the 1912 conference once again asserted its importance, and the SOS-signal became a universal signal for all radiotelegraphic companies and all countries.

Innovation in the regulations on distress communication was not only about sending the call for rescue, but also about receiving it. As the *Titanic* case showed, the crux for maritime disasters was the ability to listen. The 1912 Convention obliged operators to take a break in their communications now and then in order to ensure that the SOS-signal would be heard. The proposal was made by the United States delegation and supported by the case of the *Titanic*:

> It is essential that the operators of large vessels keep a constant watch so as not to leave calls for help unanswered (. . .) If this temporary cessation of transmissions is not imposed, it will be difficult to compel operators to listen in. This eventuality occurred with the loss of '*Titanic*'. A station operating aboard a large vessel, which could likely have saved all the passengers on the sinking vessel, did not receive the distress calls because it was constantly busy for four hours receiving messages from a long-range station. Every effort must be made to prevent the recurrence of such deplorable events and the Delegation of the United

> States makes it a compelling duty not to return to America without having made an effort to obtain from the Conference regulations to achieve this result.[12]

The proposal was generally welcomed by other delegates, but countries attempted to limit the putative radio silence as much as possible. They feared that such a measure would considerably increase the operating costs of the stations, which was an 'unfortunate consequence from the point of view of the general interest.'[13]

At the end of these discussions, the Convention placed the stations under an obligation to suspend transmission at the end of each period of 15 minutes and remain silent during a period of three minutes before continuing transmission so as to 'keep watch on the wave-length of 600 metres.'[14]

Thus, the Convention codified when radio silence was required for the purposes of listening out for distress calls, which meant a particular organization of radio ether and the creation of what today would be called a 'spectrum hole.' By definition, 'a spectrum hole is a band of frequencies assigned to a primary user, but, at a particular time and specific geographic location, the band is not being utilized by that user.'[15] Today, spectrum holes are extremely useful and important for the development of communication networks. The radio spectrum is overflooded with content, and any 'gaps' in transmission or 'holes' in the licensed part of the spectrum are therefore being constantly re-used and re-utilized. For instance, one of the innovations of the 5G networks is actually the capacity to reuse available spectrum holes in order to transmit mobile communications. In 1912, the radio spectrum was not as packed with information, however, there was already a need to squeeze in a particular content in order to make it heard – an emergency call. The idea of dropping any communication for a couple of minutes and only listening in was highly innovative, and indeed had a creative impact on the future use of radio. Moreover, the regulation stated that, if any station should hear the distress signal, it would be obliged to suspend all correspondence and not resume until the communication consequent upon the call for help was ended.[16]

Third, but not least important, the regulations embodied a highly transnational understanding of radio communication. The SOS-signal was, for instance, one of many codes of the universal radio language that facilitated transnational

12 ITU Archives, *Documents de La Conférence Radiotélégraphique Internationale (Londres, 1912)*, 224–25.

13 Ibid., 225.

14 ITU Archives, *International Radiotelegraph Convention (London, 1912)* (1912), 199.

15 Simon Haykin, "Cognitive Radio: Brain-Empowered Wireless Communications," *IEEE Journal on Selected Areas in Communications* 23, no. 2 (2005): 201, https://doi.org/10.1049/ir:20050504.

16 ITU Archives, *International Radiotelegraph Convention (London, 1912)*, 196.

communication; it was a code used beyond any national language. This international understanding of radiotelegraphy also implied uniting nations to share knowledge, provide support, and attain peace. In fact, discourse about transnational projects often had a somewhat utopian character. The conference of 1912 highlighted that radiotelegraphy would 'contribute to the extent and multiplicity of relations among nations, towards the *greater good of mankind*.'[17] Moreover, the new regulations promoted the development of various international projects that aimed to use radiotelegraphy across borders. The ideas behind such projects had been conceived in the 1900s, but their realization was greatly hastened with the need to use radiotelegraphy for transnational rescue and support (see more on those projects on p. 74).

These three features (a tool for rescue, better control over spectrum, and transnational collaborations) were relatively new aspects of the political organization of radio communication. It is interesting to note that the regulations also asserted many times that radio could be more useful when used as a one-to-many communication. Radio operators were obliged to listen to broadcasts, and the SOS signal was explicitly designed to be spread and understood by as large an audience as possible. Thus, radiotelegraphy began using its broadcasting character more widely, therefore turning its ability to radiate signals around into an important feature. Broadcasting, which had previously been considered a 'critical flaw' of wireless,[18] became an advantage in the early 1910s.

Overall, this new transnational conception of radio brought new and exciting perspectives into communication, which had previously been dependent on national decisions. It also opened up new possibilities and fostered development of new projects, even though originally the delegates had actually feared imposing any restrictions on radio. The conference of 1912 opened with the note that 'it is not to be that a Convention, however foresightful, may be prepared for all eventualities, or even may anticipate the developments of the application of a young and vigorous science.'[19] The delegates also tried to leave space for future inventions, as there was no guarantee that 'another process will not be born tomorrow.'[20] And, indeed, this feeling of something innovative and exciting at the door has been fully justified in the years that followed: the economic organization of

17 ITU Archives, *Documents de La Conférence Radiotélégraphique Internationale (Londres, 1912)*, 45; italics added.

18 Balbi, "Wireless's "Critical Flaw": The Marconi Company, Corporation Mentalities, and the Broadcasting Option."

19 ITU Archives, *Documents de La Conférence Radiotélégraphique Internationale (Londres, 1912)*, 101.

20 Ibid., 147.

wireless industry, the applications of radio technology, and the users of this communication tool themselves have been changing rapidly in the course of a general advance towards a better connected, transnational world of communication.

The political economy of radiotelegraphy: 'born global' firms and collaborations

The market for radiotelegraphy was in fact already intensely transnational long before the political decisions of 1912 had been taken. European governments supported radiotelegraph projects that helped to facilitate information flows with the colonies or to expand into and influence other independent countries, with the idea of establishing communication with colonies common within the imperial framework. As Potter has plausibly argued, even though media historians in the last few decades have tended to set aside ideas about cultural imperialism and instead write about cultural hybridity, imperial connections were still an essential part of the development of communication networks.[21]

Before World War I, Marconi, with the support of the British government, had tried to build the Imperial Chain.[22] The French government likewise decided to establish an intercolonial network of radiotelegraphic stations so it could enhance communication with the metropolis and help ships out at sea with navigation.[23] Telefunken did the same, signing a contract with the German government for long-distance wireless stations in certain parts of their dominions, including the Indian seas, the Australian seas, and the Southern Pacific,[24] while the Russian government tried to establish a network with the help of Siemens and Halske.[25] The Americans also sought to establish a network that would allow communication between Washington, San Francisco, Colon, Hawaii, Samoa, Guam, and Luçon.[26] All these projects were supposed to make communication between the various portions of the empires in question more affordable in peacetime; for strategic reasons in case of war, they would also serve as an alternative to cable, which is liable to be cut. Therefore, even before the war, there were great advances in radio technologies in

21 Simon J. Potter, *Broadcasting Empire: The BBC and the British World, 1922–1970* (Oxford: Oxford University Press, 2012), 239–40.

22 Marconi Archive, "Documents Relating to the Imperial Wireless Scheme, 1911–15" (University of Oxford, Bodleian Library, MS. Marconi 219, 1915).

23 "Réseau Intercolonial Français de Télégraphie sans Fil," *Journal Télégraphique* 3 (1912): 72.

24 "The Marconi Agreement," *The Times*, August 8, 1912.

25 "Développement Du Réseau Radiotélégraphique Russe," *Journal Télégraphique* 4 (1912): 95.

26 "Etats-Unis d'Amérique," *Journal Télégraphique* 4 (1912): 192.

various extra-European countries, which began with the initiatives of European governments and deploying the resources of various different companies.

The politics surrounding wireless had a Eurocentric character not only in light of the European nature of the technology but also because European politicians had accumulated power over the radiotelegraph networks in their hands. The colonies, although geographically crucial for the wireless network, did not have any voting power. Moreover, the empires sought to restrict the powers of their colonies. A representative example is the Egyptian government's bid to participate in the international time conference, overridden as it was by the British Empire. In 1913, the Egyptian government received an invitation to send a delegate to the conference, as it occupied a valuable geographical position. The government was ready to accept the invitation and considered becoming a party to the proposed Convention on the use of Wireless Telegraphy for Time Signals, however, in their correspondence on this issue, British officials expressed doubts about giving the Egyptian government such an exclusive right to participate in the international convention and decided to keep it out of the international debates. Instead of allowing the construction of a powerful wireless station controlled by the Egyptian government, it was suggested that the Imperial Wireless Station controlled by the British government be used for the time network.[27] An independent Egyptian representative was therefore no longer needed, as all the issues regarding the inclusion of Egypt within the international network conveniently rested on the shoulders of the British delegates.

Yet the goals of the expanding radiotelegraph networks were still highly transnational. The expansion was powered not only by imperial but also by commercial interests that determined a thoroughly transnational organization of the industry. Radiotelegraph was developing through the spread of global networks, connecting remote countries and even continents. From the point of view of the political economy of communication, radiotelegraphy was experiencing globalization before the globalization era. The work of wireless companies transcended any national label, such as 'British,' 'German,' or anything else. Just as today's Google or Facebook are American companies with their different offices scattered around the globe, the wireless companies of the beginning of the twentieth century were also 'born global' firms. It is thus not only the digital era that has stimulated the rise of the 'born global,' for there has in fact been a global process of increasing globalization that dates back to at least the nineteenth century. The transnational exchange of goods was a central concern of the trading companies

27 BT Archives, "POST 30/2576B. Transmission by Wireless Telegraphy of Time Signals and Weather Reports (Change to 1913–1914)."

of that period; various scholars underline that the histories of the Halifax and Bermudas Company, the Direct West India Company, or the Haiti Company, become incomprehensible once squeezed into a nationalistic framework.[28] Similarly, the radiotelegraph companies were operating across national borders and were dealing with the transnational exchange of information. Spreading the networks of the radiotelegraph in general required establishing new stations and offices around the globe. Many media industry players were constrained to develop as global actors, with a range of different branches, registration in different countries, and operations under various legislations. The radiotelegraph is a representative case of such a communication network, one that transcended national borders and made companies operate in relation to a global endeavor. The existence of such media players also serves to show that radio was transnational by its very nature.

The correspondence register of the ITU indicates that some requests and communication took place independently, and from individual branches of the same company. Thus, from 1912–1914, the Marconi Company communicated with the ITU from London, Rome, New York, Buenos Aires, and Paris. The Eastern Telegraph Company kept in contact from London and Madrid, while Siemens was communicating from Woolwich, Kent, and Berlin. Telefunken, for its part, had offices in Berlin, New York, and Newton St Loe, and the Commercial Cable Co. communicated from the Hague and New York. These are just some examples from the vastly spread network of the ITU, but yet others could be cited.[29] This data demonstrates that companies functioned way beyond national frameworks; moreover, their branches were not as aligned, and they might ask ITU for the same piece of information independently. The extreme geographical distribution of the companies' offices also points to their relative independence, as those offices were allowed to communicate directly with the ITU and were not simply obliged to follow the headquarters' lead.

Moreover, the 1910s are notable also for the mitigation of disagreements between large European corporations. As discussed previously in Chapter 1, technodiplomacy was a point of contention between the various companies, and the Marconi Company, in particular, was engaged in a fierce turf war over the markets in different European countries. However, to render the communication net-

28 Jock Given, "Born Global, Made Local: Multinational Enterprise and Australia's Early Wireless Industry," *Australian Economic History Review* 57, no. 2 (2017): 158–93; Winseck and Pike, *Communication and Empire: Media, Markets, and Globalization, 1860–1930*.

29 ITU Archives, *Registres de Correspondance: Radiotélégraphie, 1912* (1912); ITU Archives, *Registres de Correspondance: Radiotélégraphie, 1913* (1913); ITU Archives, *Registres de Correspondance: Radiotélégraphie, 1914* (1914).

works coherent, to sustain them, and to spread the information efficiently, the great economic powers had to form alliances, which brought new companies to life. The existence of these collaborations demonstrates that the economic development of radiotelegraphy in the early twentieth century was not explicitly shaped by national interests but instead by corporate interest and capitalism, which 'shaped the world system' as well.[30]

One of the most noteworthy examples is the establishment in 1911 of the Deutsche Betriebsgesellschaft für drahtlose Telegraphie (Debeg), which was 55% owned by Telefunken and 45% by Marconi.[31] The establishment of this company eased the tension between the two rival corporations. With its founding, ships were able to intercommunicate regardless of the type of radiotelegraph system employed on board, and the ITU noted the decision in favor of this international collaboration 'with satisfaction.'[32] The Debeg's degree of communication with the ITU and its participation in this international diplomacy had increased by then, especially after the *Titanic* tragedy, which revealed the importance of international collaboration to the world. The 1912 regulations enhanced that transnational economy of radiotelegraphy yet further, with the two companies agreeing to exchange patents in 1912.[33] In addition to Debeg, another collaboration was also remarkable, when, in 1913, the Marconi Company and Telefunken established the Société Anonyme de Télégraphe sans Fils (SA TSF) in Belgium.[34]

These were collaborations across different imperial systems and beyond national frontiers. Every patent war can therefore be seen not only as a struggle and confrontation between companies but also as the diffusion and accumulation of knowledge and technologies in the market, irrespective of who the patent holder was.

Radiotelegraph networking beyond nations

Apart from those important collaborations between different, competing companies and nations, the development of the international projects, thanks to the

30 Winseck and Pike, "Communication and Empire. Media Markets, Power and Globalization, 1860–1910," 17.

31 Anton A. Huurdeman, *The Worldwide History of Telecommunications* (Hoboken: John Wiley & Sons, 2003), 273.

32 "Revue Télégraphique de 1911," *Journal Télégraphique* 1 (1912): 3.

33 Tworek, "How Not to Build a World Wireless Network: German–British Rivalry and Visions of Global Communications in the Early Twentieth Century," 213.

34 Headrick, *The Invisible Weapon: Telecommunications and International Politics, 1851–1945*.

shared efforts of various actors in the wireless industry, is even more noteworthy. These projects were of crucial importance to the imaginary of radiotelegraphy as a transnational technology, as they had been designed to increase interconnectedness in the world, which they envisioned as a global communication space.

Time signals network

For centuries, time had relied on national and even local infrastructures, and there was no proper way to determine the 'right' time.[35] With the advent of wireless telegraphy, many countries began broadcasting time signals by radio. The time signal, as today, took the form of an auditory message with a series of pips sounding at a particular hour. The first time signals from the Eiffel Tower were sent via wireless, presumably in 1908. Scholars frequently saw the spread of the time signals by wireless as a proliferation of universal time. In *The Condition of Postmodernity*, Harvey wrote: 'The first radio signal was beamed around the world from the Eiffel tower, thus emphasizing the capacity to collapse time into the simultaneity of an instant in universal public time. (. . .) Public time was becoming more homogeneous and universal across space.'[36]

This observation could have been entirely correct had the time signals not been organized nationally. The advent of wireless did indeed foster a radically novel understanding of time as an international public good, but it was a result of a complex process. Recent studies have shown that the proliferation of universal time was a long, drawn-out process and could not be achieved simply through the introduction of the technical means of spreading it by radio.[37] Even though radio easily crossed national borders, radiotelegraphic stations were at first spreading national time only. In Europe it meant that there was a difference of several minutes in time signals coming from places like France and Germany and, therefore, the time in question was not entirely homogeneous. Moreover, the radiotelegraph messages of time signals themselves from different countries also varied in national languages and had differently formatted combinations of 'beams.' The Greenwich observatory in London, which did not have its own wireless transmit-

35 Stephen Kern, *The Culture of Time and Space, 1880–1918: With a New Preface* (Cambridge (MA), London: Harvard University Press, 2003); Markus Krajewski, *World Projects: Global Information before World War I* (Minneapolis: University of Minnesota Press, 2014).
36 David Harvey, *The Condition of Postmodernity: An Enquiry into the Origins of Cultural Change* (Cambridge (MA), Oxford: Blackwell, 1989), 226.
37 Vanessa Ogle, *The Global Transformation of Time: 1870–1950* (Cambridge (MA), London: Harvard University Press, 2015).

ter, was put in the unusual position of receiving, and not giving, the signals. It received wireless time signals from France and Germany and noted the great difference between their formats, as could be seen in this peculiar quote: 'The Eiffel Tower tuning signals resemble the crumpling of tissue paper; those from Norddeich are more like [the] squealing of a rabbit.'[38]

The homogeneity and universality of time across space, thus, has not been a gradual and natural development of wireless time signals. Instead, it was achieved through complex political negotiations and social applications, and 1912 was again a critical watershed in this process, with the aforementioned Radiotelegraph Conference in London paving the way. In particular, the conference asserted that rapid and safe navigation depended on the ship's chronometer, which could be rendered more precise were it to use radio time signals. For a better organization of the reception of time signals, it allocated for them particular frequency bands and restricted them to only 10 minutes a day at a particular time to avoid disturbances.[39] This scheduling of radio time signals could be interpreted as the anticipation of radio broadcasting, and, moreover, as a transnational phenomenon.

Further progress in using a radiotelegraph for time signals was made at the first International Time Conference in 1912, held in Paris on the initiative of the Bureau of Longitudes. This conference gathered delegates from 16 states, members of the Bureau of Longitudes, and other guests, including Gustave Eiffel,[40] with the main topic the introduction and establishment of international time, a system that would unify time in different countries and even on different continents. The delegates agreed to sustain a transnational network of radiotelegraphic stations that would transmit time signals at a given hour, with the new organization, the Bureau International de l'Heure, supposed to coordinate its work.[41] Each radio station in the network was assigned a specific hour at which it would transmit the signal, which meant that countries were united in their efforts to send this international time, so Paris would send it at midnight; at 2 a.m., it would be broadcast by San Fernando in Brazil and at 3 a.m. it would be bedtime for Arlington in the US, and so on (Table 2).

38 "Time Signals," *The Marconigraph* November (1912): 315.

39 "Article 45," in *International Radiotelegraph Convention* (London: His Majesty's Stationery Office, 1912), 206.

40 Bureau des longitudes, *Conférence Internationale de l'Heure* (Paris: Gauthier-Villars, 1912).

41 "Scientific Time Signals," *The Wireless World* (August 1913): 299.

Table 2: List of stations transmitting international time signals and the hour of the time signal, 1912.[42]

	Greenwich civil time
Paris	0 (midnight)
San Fernando (Brazil)	2
Arlington (USA)	3
Manilla	4 (provisionally)
Mogadishu (Italian Somaliland)	4
Timbuktu	6
Paris	10
Norddeich-Wilhelmshaven	12 (midday)
San Fernando (Brazil)	16
Arlington (USA)	17
Massowah (Eritrea)	18
San Francisco	20
Norddeich-Wilhelmshaven	22

The stations did not only agree to transmit the exact time[43] at a particular hour but also changed the structure of their auditory message according to the international system. The duration of the full message was three minutes, and it was transmitted three minutes before the full hour, with each dash prescribed to the last full second. To avoid mistakes of duration and spacing, the Conference advised that the signals be made automatically and not manually.[44] These signals were referred to as 'scientific time signals.'[45]

The symbol of this international time was the Eiffel Tower, as a pioneer and the paramount example of a radiotelegraph station, so the participation of Gustav Eiffel in the Time conference was not surprising. The Eiffel Tower was envisioned

42 William J.S. Lockyer, "International Time and Weather Radio-Telegraphic Signals," in *Directory of Amateur Wireless Stations in the United Kingdom Licensed for Experimental Purposes by the Postmaster-General*, ed. A.W. Gamage (London: A.W. Gamage, Ltd., 1913), 3–6.

43 An agreed upon "exact time" (l'heure définitive) now goes under the name of Universal Time (Coordinated), UTC. It is an approximation to the time of the clock stationed at the Greenwich Meridian.

44 Bureau des longitudes, *Wireless Time Signals. Radio-Telegraphic Time and Weather Signals Transmitted from the Eiffel Tower, and Their Reception [Authorised Translation]* (London: E. & F. N. Spon, Ltd, 1915), 38.

45 "Scientific Time Signals."

as a symbol of the global communication space; the illustrations from 1914 presented it as a high tower with the globe in the background, capable of sending out signals to different lands across various oceans (see Figure 12).

Figure 12: Eiffel Tower reaches the globe with radiotelegraphy, 1914.[46]

The wireless time signals were received in different parts of Europe and beyond (see Figure 13). In 1914, the naval observatory in Washington (Arlington Station)

46 "Le Vingt-Cinq Ans de Tour Eiffel," *Le Petit Journal. Supplément Du Dimanche*, April 19, 1914.

Figure 13: Stations included into the international time signals network on the map. The red labels indicate transmission of time signal twice a day (author's elaboration on Table 2).

began regularly receiving time signals from the Paris Observatory to measure the difference in longitude between Paris and Washington by the velocity of the spread of the radio signal over the intervening distance.[47] Remarkably, the British Empire did not send any time signals, although the Greenwich time was at the basis of this project and one British delegate took an active part in the conference; some countries likewise refused to participate in this transnational network. For instance, on September 1, 1912, Japan also started transmitting time signals from Choshi on the eastern shore but refused to adopt the international scheme.[48]

Furthermore, the transmission of wireless time signals was important not only as a political project, but by virtue of their proliferation inside real houses. Magazines and journals published instructions on getting the time signals[49] and noted the 'ease' of getting them.[50] The time signals proliferated both within the mansions of aristocrats or the wealthy and among radio amateurs, changing the social value of the message itself – from being a purely infrastructural content to a means of entertainment. For instance, aristocratic families in the 1910s gathered around a radio receiver to receive a time signal – both to be synchronized among themselves but also to participate in this international entertainment.[51] Radio amateurs were also keen to catch those signals, and managed to build their stations from 'odds and ends' for that very purpose.

The spread of these accessible time signals contributed to what McCrossen calls the 'end of the public clock era,'[52] because wireless time signals delivered time directly to private houses, making public clocks obsolete. This idea of ownership and direct involvement in the process of getting a time signal could best be captured with the concept of 'liveness,' which has been researched more in relation to television studies and less with regard to radio. As Bourdon puts it, liveness is not just a technical phenomenon, but also involves a user's belief in live broadcasting.[53] The time signals were a compelling instance of such a belief in liveness, as the time signals made sense only if the listener believed that the clock

---

47 "Washington Gets Paris Time," January The Wireless Age § (1914).

48 Lockyer, "International Time and Weather Radio-Telegraphic Signals."

49 Ibid.

50 "Time Signals."

51 "Illustrated London News: Greenwich Time from Paris," no. 3876 (2017): 3876; Maria Rikitianskaia, Gabriele Balbi, and Katharina Lobinger, "The Mediatization of the Air. Wireless Telegraphy and the Origins of a Transnational Space of Communication, 1900–1910s," *Journal of Communication* 68, no. 4 (2018): 758–79, https://doi.org//10.1093/joc/jqy030.

52 Alexis McCrossen, *Marking Modern Times. A History of Clocks, Watches, and Other Timekeepers in American Life* (Chicago and London: The University of Chicago Press, 2013), 25.

53 Jérôme Bourdon, "Live Television Is Still Alive: On Television as an Unfulfilled Promise," *Media, Culture and Society* 22, no. 5 (2000): 531–56, https://doi.org/10.1177/016344300022005001.

striking in Paris thousands of kilometres away was sounding at that exact moment and not even a minute later.

Overall, time signals represented a transnational broadcasting project. The time signals addressed a transnational audience and arose from a coherent network of radio stations, comprising therefore an utterly transnational project. Even though time signals were dot-and-dash messages, they were intended to reach the broad audience, thus to use radiotelegraphy as a one-to-many communication tool. They provided a sense of immediate connection with remote events, restructuring time and space.

Meteorological reports

Another transnational project, which often went along with time signals, involved the creation of a transnational network for meteorology. The dissemination of regular weather reports from shore stations was intended to enhance safety at sea. Similarly to the time signals, the weather reports differed from country to country and were broadcast at different times. The delegates at the 1912 Conference on Radiotelegraphy in London and the 1912 International Time Conference considered the importance of weather reports at sea and, similarly to the time signals, made decisions to organize the system clearly. The Radiotelegraph Conference allocated frequency bands for meteorological reports, while the International Time Conference agreed to use the established network of time-signal radio stations for transmitting weather reports.

Meteorological studies are inherently transnational phenomena and consequently require the international coordination of scholars and experts. The meteorological and seismological communities were united in a large international network long before radiotelegraphy and in fact were among the first to employ intensively global communication networks,[54] with the radiotelegraph one of the many steps forward in the consolidation of this international network.[55] The weather reports were broadcast to ships and other stations. The stations, especially those on board ship, could in turn warn about approaching storms or any natural cataclysms. Time signals, which were picked up regularly by meteorological stations, helped to more precisely calibrate the seismographs. Seismologists

54 Paul N. Edwards, "Meteorology as Infrastructural Globalism," *Osiris* 21, no. 1 (2006): 229–50, https://doi.org/10.1086/507143; Paul N Edwards, *A Vast Machine: Computer Models, Climate Data, and the Politics of Global Warming* (Cambridge (MA), London: The MIT Press, 2010).
55 Juha Herkman, "Introduction: Intermediality as a Theory and Methodology," in *Intermediality and Media Change* (Tampere: Tampere University Press, 2012), 10–29.

and meteorologists also exchanged information by radio waves, along with other means of communication. Furthermore, the meteorologists interpreted interference in radio messages as a sign of atmospheric intervention. The time signals themselves were also used to calculate longitude, the earth's shape, and the direction of the wind, calculations made possible by comparing time signals sent by different radio stations.[56] Thus, radiotelegraphy played an important role in media convergence in the meteorological and seismological fields.[57]

Meteorological information was transmitted shortly after the time signals. This seemingly simple solution in scheduling weather reports was actually the result of earlier, lengthy discussions. In 1908, French scientists discussed the possibility of organizing a coherent meteorological service.[58] Regular broadcasts of weather reports and warnings for ships were proposed, along with spaced-out time signals, enabling ships to calculate their longitude at sea and determine their geographical position.

There was one important difference between this 1908 proposal and the 1912 realization of the project. The 1908 discussion concerned both the interferences that arose through the use of radio devices and natural interferences; French hydrographer and engineer Anatole Bouquet de La Grye suggested weather reports be sent at midnight to prevent the radiotelegraph waves being disturbed by solar rays.[59] The actual scheduling of the meteorological transmissions in 1912 was quite different; it considered the convenience of radio use for both the audience and the staff rather than physical interferences alone. It was also discovered that the meteorological and seismological offices were not willing to work night shifts, even though midnight transmission guaranteed a better signal.[60] In 1912, the transmissions were therefore scheduled for a more congenial time, thus meeting the audience's expectations. This schedule illustrates how radio transmissions created the audience; at a particular hour, the operators set their devices for the time, weather, and news reception, therefore remaining silent when on air.

Proper scheduling allowed meteorological information to reach specific audiences, making radio an integral part of the listeners' daily lives.[61] Hamilton

56 A. W. Gamage, "Directory of Amateur Wireless Stations in the United Kingdom Licensed for Experimental Purposes by the Postmaster-General" (1913), 4–6.

57 Rikitianskaia, Balbi, and Lobinger, "The Mediatization of the Air. Wireless Telegraphy and the Origins of a Transnational Space of Communication, 1900–1910s."

58 "Application de La Radiotélégraphie. Avertissements Météorologiques," *Journal Télégraphique* 10 (1908): 324.

59 Ibid.

60 "Il Telegrafo Senza Fili e La Meteorologia," *Rivista Delle Comunicazioni* (1911): 983–85.

61 Bob Lochte, "U.S. Public Radio: What Is It – and for Whom?," in *More than a Music Box: Radio Cultures and Communities in a Multi-Media World* (1996), 39–40.

underscores that in contrast to the one-to-one wireless telegraph, the social concept of broadcasting presupposes a lack of reply, as a response is 'wholly and simply irrelevant.'[62] In this sense, these meteorology reports and time signals were indeed examples of broadcasting, as their scheduling did not presuppose any response. This was a simultaneous experience that was shared across the world at a given hour and that brought order into the chaos of radio communication. It was not only regular radio transmissions that were created by the inauguration of meteorology and time signal networks but also the process of, in Bob Lochte's words, 'producing audiences.' 'Producing audiences' is the most important task of radio, and in order to achieve it, radio employs the radio programming that helps to reach certain audiences, making it a part of the listeners' daily lives.[63] It is important to underline, moreover, that the decision to schedule these transmissions regularly in different parts of the world highlights the transnational character of radio, as it targeted a truly transnational audience.

Some meteorologists and seismologists flourished professionally by making use of the radiotelegraph. Most importantly, these figures also had a transnational significance. For instance, Father Guido Alfani (1876–1940), who was from Florence, became known as 'the father of earthquakes.' Alfani owed his fame to his having accurately predicted several earthquakes on the basis of exact times received from the Eiffel Tower in Paris, which were crucial to the calibration of seismographs.[64] He regularly requested information from the ITU about radio[65] and was one of the actors who shaped radio development. Another example is Albin Belar from Ljubljana, who founded the first modern seismological station in Europe, which remained in operation up until 1919. Belar was invited to numerous seismological congresses and played a role in the establishment of new, modern seismological stations in the Austro-Hungarian Empire.[66] He recognized

62 James F. Hamilton, "Excavating Concepts of Broadcasting: Developing a Method of Cultural Research Using Digitized Historical Periodicals," *Digital Journalism* 6, no. 9 (2018): 11, https://doi.org/10.1080/21670811.2018.1481762.

63 Lochte, "U.S. Public Radio: What Is It – and for Whom?," 39–40. See more on the importance of creating audiences on the historical case of Canadian radio in Anne F. MacLennan, "Learning to Listen: Developing the Canadian Radio Audience in the 1930s," *Journal of Radio and Audio Media* 20, no. 2 (2013): 311–26, https://doi.org/10.1080/19376529.2013.825534.

64 Guido Padre Alfani, "Estratto: L'Osservatorio Ximeniano e Il Suo Materiale Scientifico: V (La Stazione Radiotelegrafica). Firenze: Tip. S. Giuseppe, 1912," *Rivista Di Fisica, Matematica e Scienze Naturali – Pisa* 13 (1912): 148.

65 ITU Archives, "D. 19. N. 691.," *Registres de Correspondance: Radiotélégraphie* (1913); ITU Archives, "Alfani. D. 19. N. 691.," in *Registres de Correspondance: Radiotélégraphie* (1913).

66 Andrej Gosar, "Centennial National and Institutional Reports: Seismology and Physics of the Earth's Interior. 79.47. Slovenia," in *International Handbook of Earthquake and Engineering Seis-*

the great potential of radio technologies for seismology, which he explained in many articles regarding the use of meteorological reports and time signals in seismology. Belar made extensive use of radio technology to obtain information required for the calibration of seismographs, and taught other seismologists this technique in his public speeches and publications.[67] He was a regular subscriber to, and correspondent with, the ITU.[68] He is also known for the construction (together with Baron Codelli) of a wireless receiver for the reception of accurate time signals to be applied in seismological research (1910) and for his launching of the first radio programs in the Slovene language.[69]

Overall, the case of meteorology clearly demonstrates that in the early 1910s, radiotelegraphy played an essential role as a communications tool that provoked transnational interactions, with its purpose to transmit relevant information to different nations. These transmissions were realized through the collaboration of different companies or national radio stations. It also easily crossed borders, being consumed on a global level or with the help of small local broadcasting stations. Radio was a product of transnational importance.

Experiments with voice and music broadcasting

The first voice transmission over radio waves was reportedly achieved in 1906. That year, Reginald Fessenden proved the possibility of wireless transmission of the human voice by his radiotelephone experiments at the Brant Rock Station in the US.[70] It was an essential step towards radio broadcasting, even though the first messages lacked intelligibility. That same year, another American, Lee de

mology. Part B: Project of the Committee on Education, International Association of Seismology and Physics of the Earth's Interior, ed. William Hung Kan Lee et al. (Amsterdam, Boston: Academic Press, 2003), 1432.

67 See, for example, Albin Belar, *Ein Erinnerungsblatt, Gewidmet Der 'Società Sismologica Italiana' Zu Ihrem 10 Jährigen Gründungsfeste* (Laibach: Kleinmayr & Fed. Bamberg, 1905).

68 ITU Archives, "D. 21. N.43. Belar," in *Registres de Correspondance: Radiotélégraphie* (1911); ITU Archives, "D. 19. N. 95. Belar," in *Registres de Correspondance: Radiotélégraphie* (1912); ITU Archives, "D. 19. N. 409. Belar," in *Registres de Correspondance: Radiotélégraphie* (1911); ITU Archives, "D. 19. N. 114. Belar," in *Registres de Correspondance: Radiotélégraphie* (1913); ITU Archives, "D. 21. N. 137. Belar," in *Registres de Correspondance: Radiotélégraphie* (1911); ITU Archives, "D. 21. N. 143. Belar," in *Registres de Correspondance: Radiotélégraphie* (1911); ITU Archives, "D. 19. N. 87. Belar," in *Registres de Correspondance: Radiotélégraphie* (1913).

69 Gosar, "Centennial National and Institutional Reports: Seismology and Physics of the Earth's Interior. 79.47. Slovenia."

70 Huurdeman, *The Worldwide History of Telecommunications*, 278.

Forest, added a third electrode to Fleming's lamp and noticed a much better reception of the waves; he had discovered an amplifier effect, which was one of the greatest inventions of the time because it allowed him to create a detector and amplifier of exceptional quality.[71] This turned out to be a key for voice transmissions because the oscillations of the triode were very regular and could easily be modulated by a microphone. De Forest patented his invention under the name 'Audion.'

The Audion and the spread of amplified sound had a genuinely transnational story. In his attempt to build a successful company and overcome obstacles caused by the patent wars, Lee de Forest travelled to France. There, he managed to broadcast several voice programs, including the famous one in 1908 from the Pointe du Raz in Brittany, which consisted of several spoken and musical sequences and was captured by the Eiffel Tower 500 km away.[72] No matter how successful his experiments were, Lee de Forest had to return to the US in 1910 without having built a prosperous and influential business. He continued his experiments in the States, one of which was the broadcast of a recital of the tenor Enrico Caruso from the Metropolitan Opera. This sensational breakthrough—a live music broadcast—was appreciated by an audience of no more than 50 or so amateurs.[73] The importance of the Audion was discovered only with time, when more and more engineers became acquainted with the new technological idea. In fact, many European specialists worked abroad and picked up ideas and technologies while away. Later, they brought those ideas back to Europe and developed them in their homelands. For instance, French engineer Paul Pichon had worked for Telefunken before 1914 and had made work trips to the US; he may perhaps have brought some of de Forests' Audions to the attention of the French authorities.[74] A patent was filed in 1915, and the invention met with great success during World War 1, with over 100,000 pieces being manufactured in France.[75]

At first, wireless voice transmission was used for making point-to-point calls and therefore competed with the electric telegraph and telephone. The use of ra-

71 Huub Wijfjes, "Spellbinding and Crooning: Sound Amplification, Radio, and Political Rhetoric in International Comparative Perspective, 1900–1945," *Technology and Culture* 55, no. 1 (2014): 148–85, https://doi.org/10.1353/tech.2014.0013.

72 René Duval, *Histoire de La Radio En France* (Paris: Alain Moreau, 1979).

73 Ibid.

74 Russell W. Burns, *Communications: An International History of the Formative Years* (IET, 2004), 404; Michel Simeon, "Les Télécommunications et La Guerre de 1914 à 1918. Avènement de La T.S.F." (ITU Library & Archives, 1989).

75 Keith R Thrower, "Technical Factors Affecting CW Radio Communication in WW1," in *Innovating in Combat. Telecommunications and intellectual property in the First World War* (Reading, 2014).

diotelegraphy was regularly called *telephonie sans fils, radiophonie* or a wireless telephone. This modality had certain advantages over cable communication; for example, submarine cables at that time did not allow telephonic communication over distances greater than 60–80 km. For this reason, Corsica and Algeria were never connected telephonically to France, however, radiotelephony made it possible to make calls from France to Corsica, and experiments to reach Algeria also took place.[76]

Apart from this point-to-point use, the first attempts to organize radio broadcasting took place on the basis of the radiotelephone. This endeavour corresponded closely with the experiments with a wired telephone or a so-called circular telephone (or telephone newspaper). The circular telephone was indeed one of those precursors to radio broadcasting, the history of which still remains little researched in media studies with few exceptions. The idea, introduced in the 1890s, was to turn the telephone into a real means of mass communication; the subscribers to the service could listen live to news, operas, comedies, and concerts via wired telephone.[77] The two most famous examples of this service in Europe were Telefon Hírmondó in Budapest, launched in 1893, which reached over 6,000 subscribers, and Araldo Telefonico in Italy, launched in 1910 with around 1,300 subscribers. The circular telephone brought entertainment, music, and news to houses over telephone lines just as later radio broadcasting did. Thus, those first experiments with broadcasts via radiotelephone in the 1910s were no more than the continuation of the circular telephone but without wires.

Various different experiments were conducted to improve radiotechnology and use music and voice transmissions. Tests took place between the radio station of the Eiffel Tower and another station of the French Administration of Post and Telegraph. The transmissions deliberately did not include music, as it was difficult to judge the quality of its transmission, but included voices reading newspapers as well as having direct conversations. Interestingly, the conversation was heard by other nearby stations, despite atmospheric disturbance, and even though there had been no intention to reach them.[78] These experiments revealed that even though the transmissions were only intended for particular stations, they already functioned as real broadcasts and reached more listeners than expected.

76 "La Téléphonie sans Fil. Conférence Faite Par M. Le Lieutenant de Vaisseau Colin à La Société Internationale Des Electriciens," *Journal Télégraphique* 1 (1910): 3–9.
77 Gabriele Balbi, *La Radio Prima Della Radio. L'araldo Telefonico e l'invenzione Del Broadcasting in Italia* (Rome: Bulzoni, 2010).
78 "La Téléphonie sans Fil. Conférence Faite Par M. Le Lieutenant de Vaisseau Colin à La Société Internationale Des Electriciens."

As music and voice transmissions required considerable power, the range of transmission was often too small to reach a large audience. However, it was still a cross-border medium and was not considered a national prerogative. Radiotelegraphy was able to carry voice transmissions, music, and entertainment for listeners in a limited area, but sometimes it was also transmitted in border zones.

One of the most notable examples of music and voice transmission in Europe was in Belgium. In 1913, Albert I, the King of Belgium, ordered a radiotelegraph link to be established between Belgium and the Belgian Congo, and assigned University of Brussels Professor Robert Goldschmidt and French engineer Raymond Braillard to prepare the construction. Initially assembled for purely political and economic reasons, the stations soon became a source of entertainment. Volunteer artistes would come to declaim poems, read their prose works, and use a phonograph to transmit music. These events became a regular activity and were held each Saturday afternoon at 5 p.m., one of the first regular broadcasting services in Europe. The stations also enjoyed the support of Queen Elisabeth of Belgium, who was very interested in and passionate about radio, and had even learned Morse Code in order to hear news directly through a galena receiver; on March 28, 1914, Braillard and Goldschmidt even organized a special concert devoted to her.[79] This program became a veritable transnational service, as these concerts were captured in Paris and other nearby cities and countries. At the beginning of the 1910s, radio broadcasting had started to spread transnationally.

Interferences and the high demand for power were the critical complications for receiving stations, as delegates at the conference held at the *Société internationale des Electriciens* in 1910 noted. In order to compete with the Morse code transnational radiotelegraph communication, music and voice transmission required further technological advances.[80] In addition, the transmitting stations experienced problems with their microphones that complicated radio broadcasting over a great distance. All existing systems, such as those of Poulsen, de Forest or Fessenden, had a microphone that was combined with the antenna. When transmitting far, the microphones could not bear an intensity greater than 0.5 or 0.7 amperes. As soon as the intensity exceeded this value, the microphone overheated, and no transmission could be conducted until it cooled down. Thus, the equipment was clearly yet to be further improved.

The so-called wireless telephone was thus making the case for radio broadcasting. This medium had been created for regular music and voice transmissions

79 Duval, *Histoire de La Radio En France*.

80 "La Téléphonie sans Fil. Conférence Faite Par M. Le Lieutenant de Vaisseau Colin à La Société Internationale Des Electriciens."

and had the potential to entertain a large audience; it attempted to do so by effortlessly crossing European borders. The greatest problem, however, was that it could not compete with radiotelegraphy for transnational reach. The fact that music and voice transmissions could be broadcast only in a limited region was perceived as a serious obstacle for this means of communication at this date.

A universal language to communicate across borders

As radiotelegraphy was spreading around the globe, there was always an intention to construct a universal language for radiotelegraphy. Morse code was wholly based on national languages and therefore was not sufficient for this purpose; it simply codified letters in the national languages in a series of short and long beeps, and therefore still could be used only within the nation.

Thus, other codes appeared that were not based on national languages. One example that has already been discussed was the introduction of the SOS signal, the international distress signal which avoided any dependence on national languages and transmitted the call for help promptly and efficiently. As previously discussed, the SOS signal is indeed one of the most successful examples of such a code, which is universally known even today and is a part of common mentality.

Several other innovations share this same aspiration to communicate across nations: Q-codes and Esperanto. These were believed to facilitate interactions among professional radio operators but were extensively used by radio amateurs as well and allowed transnational communication.

The radiotelegraph employed the system of Q-codes. Q-codes were (and still are) a standardized list of three-letter codes starting with the letter 'Q,' where each code equals some particular message. The letter 'Q' is believed to be chosen as a starting point because in most languages it calls for the use of the letter 'U' afterwards. Thus, the operator can easily deduce: if no 'U' follows, this is evidently a code. For instance, the 'QRA' code represents a question: 'What is the name of your station (vessel)'? The existence and wide usage of Q-codes facilitated communication across radio waves, with the first references to this system found as early as 1906 when all radio operators were advised to follow the International Code of Signals.[81] A more intense proliferation of these Q-codes, however, occurred in the 1910s when the stipulation that Q-codes be used became obligatory. The Convention of 1912 listed Q-codes in the appendix to specify which codes

81 Le Département des Postes de l'Empire d'Allemagne, *Documents de La Conférence Radiotélégraphique Internationale (Berlin, 1906).*

were included in the International Code of Signals;[82] to avoid confusion, no country was ever issued an ITU prefix starting with 'Q.' Stations in Qatar today, for example, have to use the prefix 'A7.' This spread of the Q-codes had major significance for transnational communication.

One example of a Q-code is the famous US radio-amateur journal *QST*, first published in 1915 and named after the Q-code for 'General call to all stations.' Other examples are 'QSO' for 'Can you communicate'? and 'QRX' for 'When will you call again'? The number of Q-codes was so extensive that it could allow radio users to transmit all basic information about themselves and their stations without sharing another common language other than Morse. These codes were used both by radio telegraphers and radio amateurs and had a particular impact on the practices of radio amateurs.

There were also other attempts to create a universal language for radiotelegraph operation. The use of radiotelegraphy as a transnational media reflected perfectly the mood of modernity that sought to unite, standardize, and facilitate communication between various nations. The most famous universal language—Esperanto—was designed to serve this purpose, however, it never achieved the same success as the Q-codes; because Q-codes were obligatory for radio operation, they were the most common universal language. Some German radio stations, for instance, tried to use Esperanto, but they did not receive any responses and had to revert to Q-codes.[83]

Professionals and amateurs in the wireless industry of the 1910s

The job market in the wireless industry: Challenges and opportunities

Most commonly, the jobs in the wireless sector in the 1900s and 1910s were taken by male operators. However, the early 1910s were notable also for the shift regarding gender issues in wireless management. The late nineteenth century saw the coining of the well-known phrase, 'cherchez la femme'; attributed to the fictional detective Monsieur Lecoq, this phrase proclaims that in every crime, the investigator's first aim should be to 'seek for a woman.' It underscored that women frequently remain invisible, but that their capacity for any work (or crime) is greatly underestimated. The same notion could easily be applied to the

82 ITU Archives, *Documents de La Conférence Radiotélégraphique Internationale (Londres, 1912)*, 539.

83 W.F. Körner, *Geschichte Des Amateurfunks: Seine Anfänge (1909–1963)* (Hamburg: FT-Verlag Rojahn and Kraft, 1963), 35.

role of women in the development of wireless; they knew how to operate radio devices and did so frequently but were rarely considered to be professional workers. Initially, every attempt at female engagement with this new technology was viewed with skepticism and even disbelief. The 1910s did, however, witness a slight change in the paradigm. The most notable shifts happened in the USA and in Australia, with Europe following later.

The general situation in the wireless job market was not too promising. Hundreds of young men and a number of young women were training hard, in order to familiarize themselves with the principles and practice of wireless telegraphy, fondly hoping that they might earn their living from this profession. However, a newspaper in Sydney noted that the profession of wireless operator was not highly paid, required certain specific skills, and had more of a romantic flavour about it than any practical application. The Australian journalist depicted the typical wireless operator in the following way:

> There is something attractive about the idea of being wireless operator on board an ocean going liner. The cables have told us of instances, generally in American waters, where vessels have been wrecked at sea and help brought to the stricken passengers through the (. . .) endeavors of the wireless operator in flashing the 'emergency' or help signal across the raging waters. These lucky gentlemen, we are creditably informed, were hugged and kissed and slobbered over by the ladies on board and their friends on land, and even bits of their uniforms were nibbled away by the fair things who wanted something to keep as a reminder of the great curly headed 'ero who saved the ship. These worthy young fellows in Sydney see in their mind's eye visions of themselves sitting in the wireless operating room upon a big liner, drawing a fat screw, dining upon the best grub and swanky, and making love upon the poop desk to the millionairesses on board who were so enamored with the awesome wireless operator that they were willing to say to him. 'Come, 'Arry, fly with me to a distant land, and let us be one.'[84]

Moreover, along with this romantic image, the author noted that the wireless schools deceived many young people as to the actual salaries. He clarified that in Australia, the salary of a wireless operator was £6 a week, and amounted to a maximum of £160 a year, which was even less than the salary of the telegraph operators on land.[85] Thus, the wireless schools were making a profit in putting on certain training courses, but the newly graduated operators experienced grave disappointment when entering the job market with their recently acquired skills; it was difficult to land a position and the salary was not great either. Overall, there were more wireless operators available than there were ships equipped with wireless.

84 "Wireless and Women," *Truth*, April 14, 1912, 11.
85 Ibid.

The situation had changed, however, with the sinking of the *Titanic*. The above diatribe appeared in the Sydney's tabloid newspaper just a few days before the *Titanic* disaster, and its author could therefore never have predicted how the job market would change. A few short months after the *Titanic* sank all ships were obliged by the Radiotelegraphic Convention of 1912 to be equipped with wireless. Already, towards the end of 1912, the new international rules were being strictly enforced. Since every passenger ship was now obliged by law to be equipped with a radiotelegraphic device and to carry two operators, shipping companies had no choice but to hire more personnel.

When hiring new staff, many companies drew attention to a previously overlooked group of radio operators, namely, women. Initially this involvement of women in the wireless industry was greeted with some skepticism, especially in European countries. Indeed, it was seen as something out of the ordinary for a woman to be competent in this sphere, and male radio amateurs expressed condescension towards them, if not contempt. The case of the disaster of the Swedish steamer *Kristianiafjord* is representative of this attitude. On July 15, 1917, the ship encountered such a dense fog that it ran ashore on the Canadian coast. The wireless operators began transmitting the call for help, and many passing steamers rushed in to effect a rescue. They rescued the passengers, but their attempts to haul the steamer out were unsuccessful. The entire operation took place thanks to the continuous efforts of the radio telegraphists, and one of them was a woman. Miss Rydjord had previous experience in wireless operating at the radio station situated in Bergen, Norway; she volunteered to stand in for a wireless operator who had been taken ill the day before the accident and proved to be entirely satisfactory. The story was re-published many times in the international press, and yet her evident competence was dismissed out of hand.[86]

The problem was that women were believed to be more highly strung and nervous at moments of stress and therefore not to be trusted with so demanding a task. For a long time to come maritime wireless telegraphy was seen as unsuited to the feminine temperament, because women were still not trusted in a crisis. The note in *The Wireless World* had asserted that 'the feminine temperament is an uncertain factor in times of emergency' because there is always 'the likelihood of her natural weakness revealing itself at a critical moment.'[87] Moreover, those advocating the exclusion of women from wireless at sea further observed that it 'would be unfair to womankind and the mercantile marine.'[88]

---

86 "A Lady Telegraphist," *The Wireless World* (October 1917): 475–77.

87 "Marconi "Amazons"", *The Wireless World* (March 1918): 841.

88 Ibid.

In the UK, in the 1910s, only a few women obtained the Postmaster General's Certificate, and the very first to do so was the wife of a naval officer;[89] Mrs. Gateshill of Newcastle-on-Tyne had made headlines when she gained a first-class Postmaster General's Certificate for knowledge of three different systems.[90] Many female candidates did their level best to find work in the wireless telegraphy field, but in vain. The formal replies of the wireless companies to applications for the position of an operator read as follows: 'No suitable appointments are open to be filled by ladies.' One Irish woman even appealed against such a refusal. She resented the fact that she had received a reply from a 'Traffic Manager' on this issue, 'as she was a lady and not a horse and cart.'[91]

Women attempted to fight this discrimination, and were for the most part successful in Australia and the US. The need to appoint more wireless operators led to the recruitment of more women, initially on land. Gessler's study revealed that the 1910s in the US saw the novel suggestion that women be appointed more often to posts in wireless telegraph work. Indeed, female radio operators were engaged in a broader conversation about women's rights and envisioned early radio as a utopian space that would renegotiate gender roles.[92] At the beginning of the 1910s, more wireless companies started to hire women, their performance having proved to be more satisfactory than that of men. The women asserted that they would civilize modes of communication via radio waves, offering as an example the shift noticeable in the language of male operators. When male operators knew they would be speaking with a woman, they would stop being 'careless with their words' and might be observed '[toning down] their language in a way that would do credit to a well regulated Bible class at Sunday school.'[93]

As time went on, women also gave proof of their competence in emergencies. In 1912, the US Commissioner of Navigation explicitly stated that women enjoyed an equal status to men as wireless operators, and that they had already shown themselves to be working well at their posts.[94] The Commissioner referred to an incident that happened at Johnstown in America, where the great dam had burst. A woman at the post office stayed at her post telegraphing to the towns down river to warn inhabitants until she herself was submerged by the flood. Following the example of government offices, some private companies also began to hire women.

89 "Wireless for Women," *The Wireless World* (March 1917): 928–29.

90 "Women and War-Time Wireless," *The Wireless World* (November 1917): 545.

91 "Ladies on Radiotelegraphic Work," *The Wireless World* (October 1917): 479.

92 Anne Gessler, ""Purifying the Upper Atmosphere": Women's Work in Early Radio, 1905–1913," *American Studies in Scandinavia* 46, no. 1 (2014): 87–102.

93 As quoted in ibid.

94 "Woman and Wireless," *Evening News*, December 4, 1912, 8.

For example, the Northern Pacific wireless station, Seattle, US, appointed two women as wireless operators on ships. It was specified, however, and without any further explanation, that women were expected to be more reliable than men.[95]

Another occupation which also required wireless skill was, surprisingly, that of servant in wealthy families, most typically in Australia or New Zealand. Female migrants heading to Australia and New Zealand sometimes managed to learn and master the skill of wireless in the course of their long journey, with this skill considered an advantage at the Labour Bureau when they were looking for a job as 'colonial housewives.'[96] The Dominions Royal Commission in 1914 issued a report that stated specifically that the skill of wireless telegraphy was particularly valuable to employers. Radio telegraphy in the 1910s was seen as a transnational medium by which to receive and send news abroad, just as later, in the 1930s, radio would be promoted as a medium that could connect Australia to the rest of the world.[97]

Apart from a serious consideration of the skills of women, there were also ironic interpretations of the involvement of female operators. To quote from an argument that featured in one such article: 'It is the most modern way of talking, and wasn't it once asserted that the surest and quickest way of spreading news was to telephone, telegraph, and tell-a-woman? So that, obviously, it must be still quicker to combine two of them.'[98]

There were also some cartoons that showed women stationed at wireless apparatuses. One depicted two women on the line to each other, while a third was listening in, using antennas on her head (Figure 14). Female engagement with radio thus occasioned not only skepticism but also derision.

Nevertheless, throughout the 1910s the wireless sector began to offer jobs for both males and females. Despite lingering skepticism towards the skills of women in the wireless industry, more and more women became involved in this profession. This shift reflects not only a change in the way that the position of women was understood, but also demonstrates the growth of the wireless industry and the need to recruit more personnel. This history of a shift in the nature of female employment is also a transnational history; it mostly started in US and Australia, countries that were greatly preoccupied with transnational and global connections, with Europe lagging a little behind because of its skepticism about the skills of women.

95 "Women for Wireless," *Daily Herald*, April 8, 1912, https://doi.org/10.1177/095574900001200206.
96 "A Novel Fact That We Announced . . .," *The Wireless World* (March 1914): 749.
97 Catherine Fisher, "World Citizens: Australian Women's Internationalist Broadcasts, 1930–1939," in *Women's History Review* (2018), 1–19, https://doi.org/10.1080/09612025.2018.1506554.
98 "Woman and Wireless."

Figure 14: Ironic image of a woman listening in (1915).[99]

The scattered community of radio amateurs

This deep attraction for the new technology also led to the appearance of a particular group of enthusiasts: radio amateurs. These amateurs are acknowledged as having played a decisive role in applying and re-inventing the technology of radio communication at the beginning of the twentieth century; in particular, Kristen Haring (2007) and Susanne Douglas (1989) have demonstrated the importance of radio hams in the United States.[100] This group has been overlooked in Europe, however, and the history of radio amateurs is typically presented through landmarks and heroes meaningful for national history.[101] Nonetheless, such amateurs did indeed constitute an important group of radio users who experienced all the downsides and advantages of radio communication, gathering around their devices in small radio clubs and meeting over the radio waves.

---

99 "When a Woman Walks through a Wireless Telephone Wave," *The Wireless Age* December (1915): 210.

100 Haring, *Ham Radio's Technical Culture*; Douglas, *Inventing American Broadcasting, 1899–1922.*

101 See J. Clarricoats, *World at Their Fingertips. The Story of Amateur Radio in the United Kingdom and a History of the Radio Society of Great Britain* (London: Radio Society of Great Britain, 1967); Clinton DeSoto, *200 Meters & Down. The Story of Amateur Radio* (West Hartford, Connecticut: American Radio Relay League, Inc., 1936); Körner, *Geschichte Des Amateurfunks.*

The influence and actions of certain radio amateurs regularly exceeded the national framework. They were evolving on the transnational landscape and picking up innovations from different fields and countries; in particular, they followed news about the international conferences, familiarized themselves with the international conventions and even communicated with the relevant international organizations. Radio amateurs never actually joined any national radio amateur associations in the 1920s and therefore typically remain outside of radio historiography. In contrast to research based on the lists of the national associations, the transnational approach used in the current book has served to identify radio enthusiasts without any institutional allegiances, using for this purpose the ITU correspondence register (see more on p. 25).

Small entrepreneurs and hobbyists played an important part in the history of radio during the 1900s and 1910s. On the one hand, they made a significant contribution to the development of radio by experimenting in their backyards and thereby acquainting neighbours with the innovation itself, as well as inventing successful applications for this technology. On the other hand, their transmissions interfered with other services, which constituted a danger to radio development and required control and organization.

The ease of obtaining equipment was the major contributory factor to the particularities of radio amateurism in the 1910s. Unlike other communication networks such as railroads, electricity, the telegraph, or the telephone, radio communication required no massive cable extensions. A simple radio device could be assembled from relatively cheap materials. For amateurs and hobbyists, radio was one of the rare affordable types of media; the first radio equipment could be built 'from odds and ends' and was described in the following way: 'If you possess a few empty jam jars, a roll or two of copper wire, a rubbish drawer and 30s [30 shillings], you can erect a wireless receiving station, and by spending another 30s you may also signal through space to anybody who has a similar installation within a radius of six miles.'[102]

This affordability of radio devices also had its drawbacks. Amateurs mostly used spark transmitters, with their basic components including a battery supply, key, induction coil, spark gap, and connections to an antenna and an earth. This construction propagates broad spectrum electromagnetic waves every time the spark gap fires; with some adjustments, spark transmitters could send signals over a considerable distance and with considerable power. Their biggest drawback was that they used an extensive range of frequencies, so that the signals would interfere with transmissions on nearby frequencies. Overall, the radio spectrum was not or-

102 "Wireless Station for 30s," *Daily Sketch*, February 6, 1913.

ganized and was rather chaotic; amateurs were the ones who contributed the most to that chaos. In 1912, an Italian delegate to the ITU conference suggested discussing the issue of amateurs and disturbances in the spectrum,[103] however, this conversation was postponed until the next meeting because it was not felt to be a matter of great urgency. The recently introduced measures to organize the spectrum were seen as sufficient to combat the interference of amateurs, especially given that there were not so very many amateurs at that time.

The first radio amateurs mainly operated on an individual basis. Most of them listened to the radio through headsets and without a loudspeaker,[104] which also contributed to the individualistic sense of radio operating experience. Many amateurs at that time operated only over short distances; the existing equipment allowed them to reach out for a maximum of 20 km, which roughly equals the coverage of a local community. After some time transmitting and picking up messages within this small area, these amateurs got to know each other better and started forming communities on local and regional levels. For instance, in 1911, Arthur F. Carter of Aylesbury, UK, set his first receiver to exchange messages with five other stations up to a distance of about ten kilometres, thereby establishing a small radio amateurs' community on waves.[105] With the increasing sophistication of radio equipment, amateurs became skilled at calling across a larger and larger distance, and the first official radio clubs later grew from city into regional associations. The emergence of these radio amateurs' communities was an entirely bottom-up process, with the early radio amateurs often randomly assigning call signs to themselves.

Scholars have interpreted this absence of national associations or international organizations as a lack of control. The radio was believed to be in a 'state of virtual anarchy,'[106] however, there was another side to the coin. The absence of control allowed radio amateurs to experiment freely without following any prescriptive rules. Radio amateurs used all frequencies, different lengths of waves, and a great variety of radio devices; they were not just interfering with other services but were also colonizing the radio spectrum. In the absence of national legislative frameworks, this appropriation of any possible radio transmission and reception had a transnational and inspiring character.

103 ITU Archives, *Documents de La Conférence Radiotélégraphique Internationale (Londres, 1912)*, 331.

104 Wijfjes, "Spellbinding and Crooning: Sound Amplification, Radio, and Political Rhetoric in International Comparative Perspective, 1900–1945."

105 Clarricoats, *World at Their Fingertips*, 13.

106 Alvin F Harlow, *Old Wires and New Waves* (New York: Arno Press and the New York Times, 1971), 467.

Furthermore, the existence of universal codes on the radio (see the Section on A universal language to communicate across borders on p. 88) also allowed radio operators to communicate beyond the boundaries of their national languages and therefore to experience a mutual influence and share one and the same culture. In addition to the official Q-codes, there were also a number of abbreviations widely used in radio transmissions. Even though most were English-based, these expressions travelled across national borders and could be found in different national contexts. For instance, up until today ham radio operators have concluded their messages with the signal '73,' meaning 'best regards.' Like the SOS signal, it is easy to remember because of its symmetrical combination of dots and dashes (--... ...--). This signal was already being used in the mid-nineteenth century to conclude messages transmitted by the wired telegraph but was soon spread around the globe through radio amateurism.[107]

When describing the formation of radio broadcasting in the US, Douglas has drawn attention to what she holds to have been the main intention of amateurs, namely, an exploration of the technical possibilities of radio, which she calls 'exploratory listening.'[108] They tested their equipment and skills by catching distant signals and collecting call letters from as many faraway stations as possible. VanCour[109] complements this definition of an early radio amateur with another factor; those amateurs who were more attracted by the content of the messages pursued engagement with particular programs in the local community.

European radio amateurs also shared some interests with their American counterparts. Their exploration of radio went along with transnational projects and networks developing in Europe and around the world. Radio amateurs exchanged and listened to predominantly six types of messages via radio: technical information, time signals, weather reports, news, voice and music transmissions, games.

The first and the most common message carried technical information. The vast majority of the Q-codes raised the topics of equipment, clarity of the signals, and technical details. To exchange technical information about the station was one of the priorities. An amateur established a connection over radio waves, asked for a postal address, and sent a postcard to say what type of equipment had been in use when the signal was heard. This is a typical example of the 'exploratory listening' to which Douglas alludes, in that it rarely transmits any other

---

107 DeSoto, *200 Meters & Down. The Story of Amateur Radio*, 92.

108 Douglas, *Inventing American Broadcasting, 1899–1922.*

109 "The Sounds of "Radio": Aesthetic Formations of 1920s American Broadcasting (PhD. Thesis)" (University of Wisconsin-Madison, 2008).

message than the message about the medium. In this case, radio is used as a medium that is the message itself.

The second type of information was time signals. As previously mentioned, since the 1912 ITU Conference, the Eiffel Tower had become the central reference point for listeners because it signalled the hour to Europe (see more on pp. 77–78). Individuals and companies all around Europe received the messages and could calculate their own time depending on the difference in the time zones. A cartoon from *The Wireless World* depicted a passionate radio amateur receiving a time signal using a bizarre contraption in his backyard (Figure 15). He used a garden rake as an antenna and a bicycle to ground his radio station: what the caption describes as 'odds and ends.' The caption also indicates the use of a 'pair of telephones,' meaning headphones, which also calls to mind the aforementioned controversy in using different names for new media, particularly headphones and telephones.

This cartoon also plays upon a paradox regarding the collective or individual experiences of listening to radio waves; even though the radio amateur is the only one with a headset, and is therefore the only one actually listening to the message (therefore being an individual listener), the process is presented as a fully collective effort and experience shared by his wife, neighbors, two cats, and a dog.

Another type of information is the weather report that usually accompanied the time signal transmissions. Initially, the main purpose of sending weather reports was to warn ships at sea about imminent storms. In the 1900s, there was some evidence of weather forecasts relayed by radio broadcasting, but the practice only became commonplace after the *Titanic* catastrophe in 1912. Weather reports were one of the methods implemented to enhance safety that simultaneously met with a wide response in the radio amateur community.

Hamilton has called the audiences created by broadcasting 'a form of sociality paradoxically collective in its pervasiveness but individualized in its experience.'[110] This idea could also be relevant for the audiences of weather forecasts and time signals, especially radio amateurs. In their receiving of the messages they were not a coherent group of amateurs and yet they were nevertheless forging a collective experience. The radio broadcasting of time signals was therefore indeed a listening practice that created radio audiences, something regarded by many researchers as the essence of the later kind of radio broadcasting.[111]

---

110 Hamilton, "Excavating Concepts of Broadcasting: Developing a Method of Cultural Research Using Digitized Historical Periodicals," 10.

111 Andrew Crisell, "General Introduction," in *More than a Music Box: Radio Cultures and Communities in a Multi-Media World* (New York and Oxford: Berghahn Books, 1996), ix.

Figure 15: A British radio amateur receiving time signals from the Eiffel Tower, 1913.[112]

An irony of the early days of radio is that, in contrast to technical, time, and weather information, there is only very patchy evidence to suggest that radio amateurs were receiving, forwarding, or transmitting news. However, news that was spread wirelessly still existed in several different ways, mainly in collaboration with governmental and military structures. Thus, Italian radio amateurs received

112 "The Experiences of an Amateur," *Wireless World* 1, no. 8 (1913): 507.

the official news bulletin from the Italian Navy,[113] and Portuguese amateurs even organized their own radio news programmes.[114]

As previously mentioned, radio amateurs had also contributed to voice and music transmissions. Some amateurs listened to the Belgian concerts, for example (more on p. 87). They also organized broadcast programs of their own, targeting a small audience. A Portuguese amateur named Fernando de Medeiros, for instance, delivered one of the first music broadcasts in 1914 by linking a phonograph to a microphone and transmitting the 'Festival of Wagner.' The listeners were three other amateurs in his local community.[115]

Finally, the last type of information was ludic in nature. The messages transmitted over the open sea were initially not very numerous, so operators developed a way of playing games of chess and draughts over the radiotelegraph to entertain themselves. The first matches were recorded as early as the 1900s between liners crossing the Atlantic, and sometimes even the passengers joined the operator and collectively participated in this activity. In coding the chess notations, the radio operators regularly picked up codes invented in the 1880s for playing chess over the electric telegraph. Overall, playing chess over the radiotelegraph was seen as having 'almost the same convenience as if the combatants sat face to face, each calling to his aid all that he has of foresight, brilliancy, and resource both in attack and defence.'[116] This passion for chess was shared both by professional operators and by amateurs, and flourished later in the 1920s. In fact, the first broadcasting schedules show that along with music, weather reports, and news, chess occupied quite an important place, especially in Germany. For example, on October 14, 1928, different programs with lectures and hints for chess players were transmitted by stations in Königsberg, Breslau, and Cologne.[117]

These types of possible messages are listed here from most common to less frequent and are representative of the radio amateurs' community in the 1910s. Furthermore, this activity also corresponds to another axis: from local to transnational. Chess was played as a type of point-to-point entertainment, starting with operators on board the ship; music and news dissemination took place mainly among local communities; the interception of long-distance signals and signals

113 Balbi and Natale, "The Double Birth of Wireless: Italian Radio Amateurs and the Interpretative Flexibility of New Media."

114 Ana Paula Silva, "From Point-to-Point to Mass Communication: The Radio in Portugal from 1898 to 1939" (Lisbon: Institute of Social Sciences, University of Lisbon, 2010).

115 Ibid.

116 "The Early Days of Wireless Telegraphy. How a Chess Match Was Played," *The Wireless World* 6 (1913): 202–3.

117 "Sunday, October 14th," *Wireless World* 10 (1928): 507.

that were supposed to be broadcast, such as weather reports and time signals; and finally, transnational technical contacts among radio amateurs, which interested them the most and which represented their transnational aspirations and the transnational nature of their community.

Amateurs played a crucial role in relaying knowledge about the new technology to a range of different social groups. In the 1910s, they were not only experimenting with the applications of the technology but were also creating an essential link between the community of experts and ordinary users. As much is evident from consideration of three different prominent figures, namely, the abovementioned Albin Belar from Ljubljana, which was then in the Austro-Hungarian Empire; Luis Cirera de Terré from Barcelona, Spain; and Abel Gody from Amboise, France. The pattern of their respective interactions with the new technology was markedly similar.

These three figures were mainly known as specialists in other fields. Albin Belar, for his part, contributed to seismology. Indeed, through his efforts Ljubljana became an important center for European seismological science, with one of the very first modern seismological observatories being built there. Luis Cirera de Terré left his mark in medicine as a specialist in electrotherapy, which he saw as an indispensable treatment.[118] Abel Gody from Amboise, France, followed in his father's footsteps and devoted himself to his work as a clocksmith.[119]

In all three instances, their engagement with radiotelegraphy arose out of a keen interest in new technologies supported by a professional necessity. Their experiments with radio were driven by the idea of using the new technology to further enhance contributions made in their own specialist fields. All of them used radiotelegraphy, but in various combinations. For instance, time signals were essential both for Abel Gody, who was directly involved in the timekeeping business, and for Albin Belar, who profited from the exact time signals in his calibrating of seismographs. Luis Cirera de Terré experimented with radio waves in medicine, as well as being one of the first radio pioneers in Spain, operating in Barcelona from 1904.

All three figures communicated with the ITU and followed the news in the international arena regarding radiotelegraphy. In doing so, each was involved in transnational communication about radio technology. In 1913, all three subscribed to the *ITU Journal* and corresponded with the ITU. Luis Cirera de Terré appears only a few times in the ITU records.[120] Gody was also acting on a transnational level; he was a loyal subscriber to the ITU *Journal Télégraphique* and ex-

118 Juan Juliá Enrich, *Radio. Historia y Técnica* (Barcelona: Marcombo Boixareu Editores, 1993), 65.
119 "Il Était Une Fois . . . Abel Gody," *T.S.F. Panorama* 14–15 (1991): 20–23.
120 ITU Archives, "D. 8. N. 69. Terré," in *Registres de Correspondance: Radiotélégraphie* (1913); ITU Archives, "D. 21. N. 322. Terré," in *Registres de Correspondance: Radiotélégraphie* (1913).

changed letters and information with this supranational institution. Albin Belar was a regular subscriber and correspondent.[121]

These three enthusiasts passed the results of their experiments with radio and their knowledge of the telecommunications regulations on to their professional communities. Albin Belar taught other seismologists to use radiotelegraphy for calibrating seismographs and obtaining meteorological information, transmitting this new knowledge in his publications and public speeches at international congresses. Luis Cirera de Terré summarized the latest information about the influence of radio on the human body in his various publications. Abel Gody also shared his knowledge with other hobbyists through correspondence and publications, such as the book *Scientific Practice for the Installation of the Reception Devices of Wireless Telegraphy*. Thanks to their work, more and more people familiarized themselves with radio technology.

In applying radiotelegraphy to various fields of professional expertise, these enthusiasts re-appropriated and reshaped radio technology. They invented new techniques for using the radio devices and even fashioned their new machines themselves. For instance, Abel Gody constructed specific devices for obtaining international time signals, in order to achieve a more convenient and user-friendly practice. Albin Belar is known, along with Baron Codelli, for inventing a wireless receiver for accurate time signals applied in seismological research (1910). Dr Luis Cirera de Terré was the first radio pioneer in Barcelona, and in 1913 he succeeded in sending a signal to Valencia using a crystal receiver. Through their publications and their designing of new devices, these amateurs relayed their knowledge of radio to their professional and local communities.

Participating in dialogue about the latest inventions in radio, these three men experimented with the technology and shared knowledge with colleagues in their own fields of professional expertise. In so doing, they conveyed knowledge about radio technology to various different social and professional groups at national and local levels. Moreover, these amateurs left their fingerprints on fields adjacent to radio, thereby showing how radio transcended not only national frameworks but also areas of professional expertise. The entanglement between medicine and communication, particularly the invention of X-rays and radiotelegraphy, remains under-researched, however, scholars have recently noted that it is an essential part of communication history.[122] The case of Dr Luis Cirera de Terré shows that there was indeed an intersection between radio and medicine. The examples of these

---

121 ITU Archives, "D. 19. N. 87. Belar"; ITU Archives, "D. 19. N. 97. Belar," in *Registres de Correspondance: Radiotélégraphie* (1913); ITU Archives, "D. 19. N. 95. Belar.".

122 Natale, "A Cosmology of Invisible Fluids: Wireless, X-Rays, and Psychical Research around 1900."

three professionals, each with a vastly different personality, demonstrate that different technological fields converged in the experiments of the amateurs, thereby influencing each other's development.

Conclusions

The sinking of the *Titanic* in 1912 was of epochal significance for the development of radio and wireless technology. The story revealed to the world that wireless technology meant that ships were not condemned to be helpless in the face of adversity, and that they could call other vessels for help. The repeated and desperate attempts by the *Titanic* to send distress signals had not sufficed; unfortunately, those ships that received the message were too far away and they managed to only rescue several dozen lives. More passengers could certainly have been saved, if more vessels had heard the message. The problem was that there were at that date very few rules for being on 'air.' The *Titanic*'s story would prove to be a turning point in the development of the wireless industry, since it led to the creation of rules but also to the enhancement and revision of the wireless technology. The tragic event, the reverberations of which were felt across the globe, revealed all the potential benefits and existing shortcomings of wireless telegraphy and by the same token taught a great lesson to humanity.

Following the case of the *Titanic*, in the 1910s, various projects on an international level began to develop. These used radiotelegraphy as a transnational medium and were based upon collaboration, united networks, and the idea of global interconnectedness. Initially radiotelegraphy was used predominantly at sea and for the purposes of entertainment and commerce, but the tragic fate of the *Titanic* revealed to the world the importance of radiotelegraphy and consequently transformed the social order. The ensuing 1912 International Conference on Radiotelegraphy concentrated on issues of radiotelegraphy as an indispensable common good shared by different nations and stipulated its clear organization and regular use.

In the 1910s, several transnational radiotelegraph networks were created. A network of radiotelegraph stations transmitting the exact time signal transnationally was inaugurated in 1912, along with a system for relaying meteorological information. Also in the early 1910s, news agencies appeared, their aim being to transmit and receive information transnationally. Projects serving to broadcast voice and music were also launched, and these reached out to a transnational audience. Interestingly, media and communication scholars rarely pay attention to time signals and meteorology, having as they did a high infrastructural content but serving thoroughly pragmatic interests. However, they nonetheless made a significant contribution to media history; they were important in the 1910s as they shaped the vision of radio, and in fact are still present today in the media environment in various different guises. In a way, this very 'technical' chapter opened up a history of transnational radio broadcasting.

The commercial companies, which were usually considered to be national actors, were also restrained and were compelled to reach out to each other. Various different international collaborations appeared to make the sea a more accessible and safe space for travel. Moreover, in the 1910s many companies also gained a foothold in different countries and colonies, functioning as 'born global firms' and therefore important transnational actors. Even the projected colonial networks, usually considered by historians to be a part of national history, could be seen as a transnational creation. The emergence of a universal language for radiotelegraphy, such as the Q-codes, also stimulated transnational exchange across borders. The transnational use of radiotelegraphy was also evident in the issue of employing women in radiotelegraph work; they could learn how to operate equipment on board a ship and therefore be in the transnational space *per se.* Women could also communicate with different operators in a distress situation without difficulty, but were nevertheless deemed to lack the skills needed for transnational work.

Radio amateurs became one of the most important social groups to benefit from these advances in radiotelegraphy in their everyday lives. Amateurs were not fully institutionalized and were therefore mainly operating on an individual basis. In the absence of a national hierarchy, this individual work paradoxically helped them to operate transnationally over the radio waves by listening to information intended for a broad audience. Their interest in the use of radiotelegraphy is particularly important because they introduced radiotelegraphy to their communities and became important mediators of this new technology.

Overall, this overview of the radio of the early 1910s indicates a more nuanced picture of the advent of one-to-many media. In that decade, the radiotelegraph was being used as a proper broadcasting tool and had already consolidated a specific audience around this media. Transmissions of technical information, such as time signals, weather reports, and international news, were the first radio broadcasting programs. They were not addressed specifically to every user but were nonetheless picked up by large numbers of listeners and had already created the first audiences.

Moreover, the radiotelegraph in the 1910s was highly transnational. International regulations, the work of companies, and technical projects all serve to illustrate the transnational nature of radiotelegraphy. The radiotelegraph was considered capable of spreading a global communication network; even the audiences for the very first radio 'programs' were transnational because these programs were intended to reach other nations. This transnational organization of radio had an explicitly Euro-centric character, as it was organized and controlled by the European marine powers.

Had there been no *Titanic,* decades would have passed before transnational radio communication entered various spheres of social life worldwide. The tragic event of 1912 stimulated the rise of international collaboration and brought to the fore a number of different new transnational projects that had been emerging. This era was a precursor of globalization for communication technologies, however, as a transnational era for wireless it was destined to be all too short-lived.

The development of radiotelegraphy through transnational communication networks could have brought an end to the 'interpretative flexibility' the radiotelegraph had experienced from 1890 to the 1900s. Very few issues for negotiation were left after the discussions at the 1912 international conferences, or in the organization of the radio spectrum. Using the terms of STS, this was a period when the technology was tending towards the closure of its interpretative flexibility, however, the period of its stabilization was rather short. Bijker has underscored that the stabilization of any technology has a processual character and may well take several years.[123] European society, however, did not have this time at its disposal; the vision of radio as a transnational media useful for international collaborations was soon challenged by the outbreak of World War I.

123 Bijker, "How Is Technology Made? – That Is the Question!".

Chapter 3
European wireless at war, 1914–1918

On June 28, 1914, Archduke Franz Ferdinand, the heir to the Austro-Hungarian throne, and his wife were shot and killed in Sarajevo by a Serbian nationalist. This assassination became the trigger that would spark the outbreak of World War I. Between 1914 and 1918, over 30 nations declared war and were divided into two opposed blocks; the Allies comprised Serbia, Russia, France, Britain, Italy, and the US, while the Central Powers included Germany, Austria-Hungary, Bulgaria, and the Ottoman Empire. The empires used resources from their colonies all over the globe, turning the war into a conflict on an unprecedented geographical scale. A single incident in southeast Europe thus grew into the most momentous and tragic war waged by the European powers.

This chapter discusses the challenges brought on by the outbreak of World War I in August 1914 and the general perception of radio as a transnational technology. First, the chapter discusses how the war necessitated the spread of radiotelegraphy and how war restrictions and the particularities of military radiotelegraphy influenced the development of radio. Then, this chapter reconsiders the development of radio, the uses to which it was put, and changes in the international arena outside of the theatres of World War I. Apart from the use of radiotelegraphy by belligerents, it also traces the changes and restrictions in the transnational radiotelegraph networks discussed in the previous chapter. Finally, it draws attention to users of technologies beyond the battlefield, namely, non-mobilized amateurs.

Military radiotelegraphic communication across the borders

The unprecedented scope and scale of World War I gave a massive boost to research into technologies of communication, with the substantial number of armies and fleets of the belligerent states increasing the need for communication in the war zone. Along with letters, pigeons, and dogs, new telecommunication technologies began to proliferate on all fronts. To command, control, and synchronize troops and forces, all armies used optical signals with portable projectors, electric telegraphy, telephones, and other media. Radiotelegraphy was perhaps the most innovative telecommunications deployed in this period, and one that would prove to be indispensable in certain situations.

Many of the warring parties built strategic radiotelegraphic connections to communicate with each other. In 1915, for instance, Spain and Austria signed an agreement to establish a radiotelegraphic service between Barcelona and Trieste.

Open Access. © 2024 the author(s), published by De Gruyter. (cc) BY-NC-ND This work is licensed under the Creative Commons Attribution-NonCommercial-NoDerivatives 4.0 International License.
https://doi.org/10.1515/9783111202327-004

Another connection was announced between Italy and Spain via Rome and La Spezia on the one side and Barcelona on the other.[1] In some cases, radiotelegraphy became an indispensable and crucial tool of communication; indeed, it had a direct impact on the course of the war, allowing the Allies to communicate with each other and synchronize their actions.

Radiotelegraphy was essential where it was physically impossible to lay a cable or to connect otherwise. Two instances of radiotelegraph use from either side are particularly noteworthy: communication between the Central Powers and Turkey and also between the Allies and Russia.

For Turkey, the radiotelegraph played a key role in its entry into the war. When hostilities commenced, Turkey seemingly maintained neutrality, however, on August 2, 1914, the country clandestinely joined the Central Powers in a secret Ottoman-German Alliance, with the initial, complex operation by which Turkey first entered the war involving the use of radiotelegraphy. On August 3, at 6 p.m., the two German cruisers *Goeben* and *Breslau*, at sea and somewhere between Sardinia and Sicily, received a radiotelegraphic message informing them that Germany had declared war on France and ordering them to sail to Constantinople. As they were a part of the German naval division under the command of Admiral Souchon, they rushed to Turkey's shores. The ships met with no obstacles at sea; Britain and its massive Mediterranean Fleet would only enter the war on August 4, after Germany's invasion of neutral Belgium. Moreover, at the Strait of Otranto, which later acquired strategic importance for the naval theatre of war, only one vessel was stationed at that time: the single light cruiser *Gloucester*, which did not pose a danger to the heavy German vessels. The German ships were in constant communication with their government, and their transmission of radiotelegraphic messages scrambled those of *Gloucester*. With this communication, the German division was able to follow the news about the German and Turkish alliance and even waited for three days at sea until the German ambassador in Constantinople negotiated to open the Dardanelles Straits. On August 10, the German division received a radiotelegraphic message that the British ships had set out in pursuit, but it managed to enter Turkish waters before being intercepted. The Entente requested explanations from Turkey regarding two German ships entering its waters; to appease these claims, Turkey answered that it had bought the two cruisers and thereupon began a purchasing procedure. The ships acquired new names: *Yavuz* and *Midilli*' both flew the Turkish flag and hired Turkish sailors.[2] Two months later, Turkey entered

1 "Revue Télégraphique de 1915," *Journal Télégraphique* 1 (1916): 1–3.

2 Maurice Larcher, "La Guerre Turque Dans La Guerre Mondiale," *Les Archives de La Grande Guerre et de l'histoire Contemporaine*, no. 44 (1921): 1562–63.

the war, attacking the Russian port of Odessa in the Black Sea with these two ships. Hence, the radiotelegraph was of crucial importance for Turkey when establishing connections with the Allies. German ships were ordered to sail to Constantinople, managed to synchronize their actions with the German ambassador, escaped British pursuit, and became one of the crucial instruments for Turkey in the war thanks to the efficient use of radiotelegraphy. In this example, radio was an indispensable communication tool; it worked when no other means of communication could be used, was mobile, and carried across great distances.

For the Allied Powers radiotelegraphy became crucial when the material disruptions of their infrastructural networks halted communications between countries. In particular, it played a vital role in establishing communication with faraway Russia over the course of the Russian Revolution in October 1917 when the Bolsheviks seized power. Because the cable lines were in enemy hands, the Russian Revolution of 1917 became one of the very first events in history to be communicated abroad by means of radio communication alone.[3]

This radiotelegraphic communication was also accompanied by interference and eavesdropping. As discussed in Chapter 1, as a channel for communicating information with a high degree of accuracy, radiotelegraphy was already held in the 1910s to be unsatisfactory. The two aforementioned cases of the German-Turkish alliance and communication between Russia and the Entente Powers likewise reveal how problematic interference could prove to be for communication. Just as the radio messages of the German ships scrambled those of the British cruiser, the message about the Russian Revolution became mixed up with other signals. The Allies did not comprehend the state of Russian politics immediately; for about two months, they remained confused, due to contradictory pieces of information coming in from Eastern Europe. The situation only became clear once the opposition had explained in detail and confirmed how things stood; General Denikin, who was heading to Saint Petersburg to defeat the Bolsheviks, reported his movements to the British by radiotelegraph and described the recent political events in Russia. In the end, he was defeated, but his messages served as the most accurate source of information for the British.[4,5] The maritime environment did

3 "Diplomacy by Wireless," *Wireless World* (October 1918): 389.

4 Marconi Archive, "Monthly Reports on News Services, 1916–21" (University of Oxford, Bodleian Library, Ms Marconi 302, 1917).

5 Like no other political elite of the time, the Bolsheviks understood the importance of radio as a medium for political messaging and the organization of everyday life. However, the first attempts to use the radiotelegraph were problematic due to interference, and many projects relating to the radiotelegraph encountered technical difficulties. Even the attempts after the revolution to introduce a new national time were not fully implemented until the 1920s.

not offer any other alternatives, and governments still appreciated radiotelegraphy for communication that offered mobility and covered long distances, notwithstanding the problems with interference.

In order to build successful communication, nations perforce had to share knowledge and practices in radiotelegraph use. Thus, the war had not only intensified the need for radiotelegraphy, but had also stimulated technology and knowledge exchange across borders. All European countries were compelled to put additional effort into the distribution and refinement of radiotelegraphic equipment, especially as a shortage of radio stations was evident immediately after the outbreak of war. To supply their armies with radio, governments mobilized all possible domestic resources, which often led to a transnational technology transfer. These cross-border knowledge and technology exchanges often tended to entail the reorganization of the companies involved.

The abovementioned case of Russia provides a telling illustration of technology transfer across borders. As previously mentioned, the Russian case was rather peculiar because the Germans had physically separated the country from its allies by cutting all the cable infrastructure, so that the sole remaining method of communication was radiotelegraphy. The shortage of radio equipment, however, was extreme; the Russian government could not even equip its Supreme Command with radio stations of sufficient power and range. The Russian manufacturers did their utmost to set up radio stations, but compared with European products, theirs were not advanced enough or convenient to use.[6] These shortcomings in the existing radio stations resulted in regular complications in issuing commands to and controlling troops, leading to the Russian Army suffering losses of personnel in combat. The failure to solve this problem with its own industries forced the Russian government in turn to seek help from its allies, especially from the Marconi Company, with which it already had relations before the war. From the middle of 1915, Russia began to procure Marconi receiving and transmission stations from Britain, along with receivers and amplifiers from France, and the War Department opened new radio stations on the outskirts of Moscow, Saint Petersburg, and in the regional capital Tver', specifically in order to communicate with the Allies.[7] In light of the high demand for radio sets on the front, at the end of 1915 the War Ministry instructed radio manufacturers to model their

---

6 Timofej Vladimirovich Alekseev, *Communication Industry of Petersburg-Leningrad for the Army and Navy in the Era of Commotions and Modernization, 1900–1945 [Industrija Sredstv Svjazi Peterburga-Leningrada Dlja Armii i Flota v Jepohu Potrjasenij i Modernizacii, 1900–1945 Gody]* (Saint-Petersburg: Peter the Great St. Petersburg Polytechnic University, 2010).

7 Vladimir Mikhailovich Leshhinskij, *On the Nature of the Wireless Telegraph [O Sushhnosti Bezprovolochnago Telegrafa]* (Tver: B.I., 1918).

own production on the lightweight British and French portable radios. When carrying out this order, the manufacturer ROBT and T developed several types of small spark radio stations for communication in artillery; the Russian development of radiotelegraphy thus had transnational significance, forming part of a transnational story of technology transfer. The Allies supplied the Russians with the equipment required, and it did indeed become an indispensable source of valuable information during the October 1917 Revolution.

To produce communications equipment, the European powers mobilized all their existing factories. In France, for instance, military units were supplied with receiving and transmitting radio stations thanks to the Société française de radio-électricité.[8] What was peculiar about this mobilization was the fact that this process involved not only national companies but foreign ones as well. As previously discussed, many radiotelegraph companies were operating globally, having set up branches around the globe in the 1910s. The nation-oriented war and the immediate imposition of severe restrictions led, however, to a drastic change in this global organization of the market; some branches during the war were trapped in foreign lands and throughout the course of the war had had to change their 'nationality,' with one illustrative example being Siemens. Although Siemens and Halske is widely known as a German electrical engineering firm, radiotelegraphy matters were handled by its British branch, called Siemens Brothers, whose head office was located in Woolwich, an area of London, with the firm producing cables, telephones, and radio apparatuses in the UK. In 1901, it was even trusted by the British Admiralty to be one of three suppliers of 54 wireless telegraphy sets made to their own specifications and system;[9] it was also a prominent actor in international discussions about radio at the conferences and in correspondence with the ITU.[10] From 1912–1915, it also conducted experiments on wireless in collaboration with the Galletti Wireless Telegraph and Telephone Company, which featured experiments between the Cuckmere and French coastal stations and the Crystal Palace.[11] When war had been declared and British men were being mobilized to fight the enemy, many German corporations based in the UK experienced

8 Caroline Ulman-Mauriat, *Naissance d'un Média. Histoire Politique de La Radio En France (1921–1931)* (Paris, Montreal: L'Harmattan, 1931), 11.

9 "The Admiralty Have Just Ordered . . .," *The Engineer* 6 (1901): 666.

10 See, for example, Siemens Brothers, "D. 19. N. 799," in *Registres de Correspondance: Radiotélégraphie* (ITU Library & Archives, 1913); Siemens Brothers, "D.4. N. 165–167," *Registres de Correspondance: Radiotélégraphie* (ITU Library & Archives, 1914); Siemens Brothers, "D. 4. N. 74," *Registres de Correspondance: Radiotélégraphie* (ITU Library & Archives, 1915).

11 BT Archives, *POST 30/2424. Experiments and Licences for Galletti Wireless Telegraph System, Part 1.* (London: BT Archives, 1915).

problems with maintaining their reputation, as customers refused to invest in companies whose main interests were German, being reluctant to use products and devices with the engraving, 'made in Germany.'[12] Even though such companies tried to convince the public that they were completely British-based, there was another problem: the risk of espionage. These and other factors stimulated the process of buying such companies, Siemens among them. Scrutiny of all that Siemens produced revealed that telephone and radio equipment was being produced to a relatively advanced specification, and could serve the British nation well,[13] and the Siemens Brothers company was sold in 1917. Throughout World War I, Siemens had lost 40 per cent of its capital. Most of its foreign assets and almost all of its patent rights had been expropriated, and the most urgent tasks during the immediate post-war years were the reorganization of the company's expanding manufacturing operations and the revitalization of its foreign business. This process of buying up foreign companies had thus facilitated a technology transfer between enemies during the war.

Overall, successful military operations using radiotelegraphy required a complex organization of radiotelegraphic services on a transnational level. One of the solutions to the problem of interference was scheduling the transmission; for example, in 1914, the international radiotelegraph channel between Russia and Entente's allies worked only during the night shift: from sunset to 10–12 a.m., there being too many disturbances in the signals during the day.[14] This attempt to schedule transmissions can be viewed as similar to the practice of scheduling weather reports in the 1910s or programming broadcasting content in the 1920s (as also previously discussed). Even though rapid military operations or revolutions could not be predicted or scheduled, this solution was effective for regular information exchange even in wartime.

---

12 Marconi Archive, "Papers Concerning the National Status of the General Electric Company, 1914–16" (University of Oxford, Bodleian Library, MS. Marconi 3152, 20–25, 1916).

13 Marconi Archive, "Correspondence and Papers Concerning the Status of Various Firms and Industries during the War, 1910–18" (University of Oxford, Bodleian Library, MS. Marconi 3163, 1916).

14 Glushhenko, *Place and Role of Radio Communication in the Modernization of Russia (1900–1917) [Mesto i Rol' Radiosvjazi v Modernizacii Rossii (1900–1917)]*, 60.

Coding and encrypting, refining the technology

There were certain peculiarities characterising radiotelegraphic communication. Coding, misinformation and espionage are the key pillars for understanding the characteristic features of military radiotelegraphy in World War I.

One pitfall of radiotelegraphy was the risk of being intercepted by the enemy, with special divisions formed to overhear the enemy's messages. To prevent eavesdropping, special coding was developed for military radiotelegraphy. Special divisions were entrusted with the task of refining their own codes and breaking the enemy's. For instance, the British recorded every single German message and then telegraphed them by landwire to the British Intelligence Headquarters at the Admiralty, where they were pieced together and decoded. At first, the British could not read the messages, but during the war, after the sinking of several German cruisers that had been using the latest German codes, the British obtained the decoding patterns.[15]

Information about the coding patterns was shared across national borders. Thus, the French intercepted German messages and managed to decipher some peculiarities in the encryption of the German radio messages, such as specific letters of the alphabet serving to designate each cavalry corps. The French shared this information with the Russian high command, and the code helped the Russians to read the radiograms intercepted from the German cruiser *Magdeburg*. These radio messages contained plans for the German fleet to go out to sea, and the Russians successfully forwarded this information to the British to warn them; when the Germans were about to strike the English fleet, they therefore suffered heavy losses in the Battle of Jutland.[16] Radiotelegraphy allowed information about the codes to be passed from one country to another: from Germany to France, Russia, and Britain. Furthermore, this massive naval battle, involving a full-scale clash of battleships, proved to be of the utmost importance, and is indeed believed to have brought the US into the war.

The practices of coding and decoding evolved also thanks to the Morse Code, which continued to be the main language deployed by radiotelegraph users. Morse Code allowed messages to be transmitted as far as possible, as it required less power and was easier to cipher. Even though music and voice transmission was already possible in the war years, the great majority of radio messages were still in

15 Headrick, *The Invisible Weapon: Telecommunications and International Politics, 1851–1945*, 162–64.

16 Ivan Terentievich Peresypkin, *Radio – a Powerful Means of Defense of the Country [Radio – Moguchee Sredstvo Oborony Strany]* (Moscow: Military Publishing House of the Ministry of Armed Forces of the USSR, 1948), 32.

Morse code. Some spoken messages were also in use; for example, on October 26, 1915, the Eiffel Tower received a spoken message from the American station of Arlington, which was about 7,000 km away. The importance of voice transmission was grasped only where fast and efficient communication was a necessity.

The capacity to use code, to be portable, and to not require material connections also made radio communication a handy instrument for espionage. As discussed in Chapter 2, during the 1910s, private experimenters and amateurs were able to operate devices, teach third parties, and communicate with each other. Most amateurs were equipped with basic receivers and spent their time listening to the codes from wherever they had originated, however, some were so intrigued that they mastered Morse Code and built transmitting sets. With these transmitting sets, they could easily send information abroad and could potentially be incriminated in espionage. The presence of sophisticated private experimenters constituted an intolerable risk for national security.

At the start of World War I, newspapers encouraged their readers to report on their neighbors if ever they should spot them with a radio.[17] The 1916 British novel *The Spy Hunter* demonstrated the typical mindset of the time by telling the story of radio amateur and ex-wireless operator Harry Nettlefield and his fiancée, Clotilde, who were chasing and cornering German spies working undercover in England, who disguised themselves as ordinary English men and women and transmitted information by wireless.[18] Any incomprehensible sounds over the radio waves aroused suspicion; for instance, in the US, local amateurs reported hearing unintelligible messages sent in cipher that seemed to point to the existence of a concealed station somewhere in the woods around Tampa, Florida. It was believed to be a place where three operatives had landed from a German cruiser and were sending coded messages; the rumor was investigated, and it was concluded that the strange coded messages originated from a boy learning Morse Code.[19] These examples show how a new medium can cause social fear, which frequently accompanies the first stages of a new technology, even engendering dystopian visions of it.[20] In this case, the stabilization process was concerned with the practices of transmitting and listening. Although previously seen as equally possible, these two modes of operating were challenged; while listening was still a normal practice for the masses, transmitting was now reserved for the authori-

17 "Wireless in the War," *The Wireless World* (September 1914): 375–82.

18 William Le Queux, *The Spy Hunter* (London: C. Arthur Pearson, Ltd., 1916).

19 Richard A. Bartlett, *The World of Ham Radio, 1901–1950: A Social History* (Jefferson, London: McFarland, 2007), 45.

20 David Hendy, *Noise: A Human History of Sound and Listening* (London: Profile Books, 2013).

ties. This change created an asymmetry in communication that would be carried on into the 1920s where radio broadcasting was concerned.

Radio devices could be easily concealed due to the immateriality and invisibility of the radio waves and the absence of a cable, but the antenna remained a very visible sign. Subsequently, the most critical task when erecting a wireless station unnoticed was to camouflage the aerial.

Figure 16: A spy using radiotelegraph on a cliff to communicate with ships, 1915.[21]

The popular mode of concealment involved using a clothes line, which became the main topic of many humoristic cartoons and anecdotes. A 1915 drawing in *The Wireless World* captured the image of a German spy who could discretely use radiotelegraphy to transmit messages to German ships; the "lady" with visible mus-

21 "The Wireless Spy," *The Wireless World* (January 1915): 646.

tache masked her antenna and all the other equipment by means of clothes lines, seemingly letting the laundry dry on the edge of a cliff, with her invisible communication with a ship depicted with a dotted line (Figure 16). As a result, clothes lines in general became objects of suspicion and attracted attention in neighborhoods where people were anxious about spies.

Radiotelegraphy and the existence of codes that marked 'us' and 'them' were also used to spread counter-propaganda. An open transmission was not only an obstacle to communication but was also a valuable instrument; misinformation, especially with the correct coding and style of radio messages, could easily mislead the enemy. In particular, imperfections in airplane navigation due to the shortcomings of the direction-finding sets were used to trap the enemy's planes. At the end of October 1917, seven German Zeppelins got lost in the fog on their way home. They tried to disclose their whereabouts by sending requests to their German base, but a French intelligence service intercepted their requests. Using a German cipher, the French gave them the course and guided the Zeppelins through the fog. However, instead of Germany, the course led to France. The Zeppelins were trapped; four were brought down by anti-aircraft artillery; two fell into the sea, and only one managed to reach the French base.[22]

Another counter-propagandistic example of radiotelegraph use concerns news services. Lasswell (1938) famously noted that while the British Navy was maintaining 'wireless silence' and did not broadcast news about its losses, the Germans were able to shock the British population by transmitting news about their victories far and wide by means of radiotelegraph.[23] Moreover, some of that news could well be fake. In April 1916, the Germans transmitted a report on the course of the war to the German Embassy in New York, claiming that 'England's industry to a considerable extent lies in ruins' after the incendiary bombs dropped by the Zeppelins; this report also pointed out that 'England's own soil has been ploughed up by the mighty explosive shells of German air squadrons.' The report was marked as sent by the British news agency, the Wireless Press. This supposed source was soon discovered to be false, as the information about the bombing was incorrect, and the Wireless Press had not transmitted it. Moreover, the report was sent in English, even though it was intended for Germans at the German embassy, with the newspapers seeing it as an attempt to spread terror among the British population, in that publications might intercept the message and relay it.[24] In 1916, the Germans

22 Peresypkin, *Radio – a Powerful Means of Defense of the Country [Radio – Moguchee Sredstvo Oborony Strany]*, 31–32.

23 Harold D. Lasswell, *Propaganda Technique in the World War* (New York: Peter Smith, 1938), 110–11.

24 "A Curious Misunderstanding," *The Wireless World* (April 1916): 32.

were also reported to have distributed through wireless 'whole pages of the most malicious and slanderous messages' about France.[25] This false news could not be refuted because the French lacked sufficient radiotelegraphic resources to compete with the Germans, however, much effort was put into the organization of such a prompt counter–propaganda response. Therefore, in this case the radiotelegraph was used as a form of mass media to target a particular audience. The drawback of open transmission had turned into a great advantage when the military discovered the possibility of spreading misinformation.

Media scholars and war historians have extensively researched radiotelegraphy as a counter-propagandistic tool during World War II, highlighting its importance in deceiving the enemy, however, these practices had been pioneered in World War I, in particular in the guise of the so-called *Funkspiel* ('Radio game'). *Funkspiel* was a counterintelligence war operation that involved capturing the enemy's radio operator or cipher for the purpose of using them later for propagating misinformation.[26]

Regarding the technical side of the wireless, the war led to considerable advances in radiotechnology, with the first important aspect of investment in equipment concerning mobility and the transportation of radio. The use of radio at sea was possible because radio was not attached to any cable grid, however, these powerful devices still remained quite bulky; to be mobile, they had to be carried by some equally powerful means of transportation. Ocean liners were one such instance, while railway engines were another, with special inventions such as the 'railophone' being introduced in the UK and Sweden to connect a moving train to the station wirelessly.[27]

Figure 17 illustrates the portability of the powerful military telegraph station. It was transported by several buggies and resembled a small camp rather than a portable device.

Considerable efforts were made throughout World War I to simplify, popularize, and multiply the military's radiotelegraphic equipment. A momentous leap regarding portability concerned the application of the radiotelegraph in aviation. The first experiments in radiotelegraphy and aviation took place in the early

25 "News [Vesti]," *Early Morning [Rannee Utro]*, 1916.

26 David Kahn, *The Codebreakers: The Comprehensive History of Secret Communication* (New York: Scribner, 1996).

27 BT Archives, "POST 30/2795C. Railophone System for Wireless Communication with Moving Trains (1910–1917)" (London: BT Archives, 1917).

Figure 17: A portable military radiotelegraph station, 1915.[28]

1910s, but the war accelerated this progress. Civil aviation in the 1910s had emphasized the need for a greater number of lights and audible signals for aircraft,[29] as radio telegraphy was then judged not to be sustainable.[30] Like ships at sea, vessels in the air were powerless to request assistance and communicate; to orient themselves in space and maintain a precise course, they had to fly over the ground or even to land, which therefore lengthened the flight,[31] or else use auditory and visual signals. The outbreak of the war did much to intensify the development of radio for planes, because military aviation could evidently benefit from the invisibility of radio communication.

However, there were several problems with putting a radiotelegraph on an airplane. The first obstacle was size; early radio devices used rather large antennas, and only a dirigible could accommodate them.[32] Second, their excessive weight affected the aircraft's aerodynamics, while lightweight radio sets were not powerful enough. These smaller devices allowed communicating at a maximum distance of 50–60 km, which was not sufficient for military aviation.[33] Third, the noise of the aircraft engine was also challenging for radio reception, since the signal was audi-

28 "The Signal Service. Wireless Telegraphy in the War," *The Wireless World* (February 1915): 719.

29 "Wireless in Air Navigation", *Wireless World* 2 (1919): 623–27.

30 "The Report of the Civil Aerial Transport Committee. Appendix A. International Convention in Regard to Aerial Navigation Drafted by the Conference Held at Paris in 1910," *Flight* (1918): 1465–79.

31 "La Télégraphie sans Fils et l'aérostation," *Journal Télégraphique* 7 (1909): 176.

32 "Military Aviation," *The Wireless World* 11 (1914): 516.

33 "An Interesting Lecture," *The Wireless World* 12 (1914): 577.

tory and earphones were not yet well designed. Finally, it was rather problematic for a pilot to communicate and manoeuvre the aircraft simultaneously; radiotelegraphy was still a combination of 'beams' in Morse Code and often required the listener to write the symbols down to remember the sequence. A wartime cartoon published in 1915 captured an incident involving a radiotelegraph operator in an airplane; he dropped a pencil while receiving a radio transmission and started volplaning,[34] however, even though dots-and-dashes transmission was a cumbersome means of communication with the ground, it was still the only way to avoid crashes, navigate the sky, synchronize plane movements, and receive weather forecasts. In this regard, radiotelegraphy also had a tangible and material effect on the construction of the first planes in terms of space and balance, as these planes were also specially constructed to include a radiotelegraph and even had a designated seat for a wireless operator next to the pilot.[35]

The first radio devices on planes were able to either only receive or only transmit information. This asymmetry in telecommunication led to the development of the observation military planes from around 1915, which used radiotelegraphy intensively. An observer from a plane could spot an enemy's position and transmit it to his army using Morse Code; there was even a specific table with a combination of Morse signs that coded for the particular positions of objects on a two-dimensional map.[36] By transmitting the enemy's position, a radiotelegraph operator on a plane could help to adjust artillery shots, however, planes could not communicate with each other; the first tests of a radiotelephone were carried out in June 1917.[37]

With developments in both technology and means of transportation, radio devices became smaller and lighter. From ships and railways to buggies and airplanes, the radio was gradually projecting the idea of communication in mobility; a later triumph would be radio in cars. The possibility to connect while retaining mobility is one of the most important topics in communications history, and these experiments with radiotelegraphy were among the first. Portability in general is one of the most frequently discussed categories in media history, as it greatly contributes to 'the transition from technician's delight to household convenience,' as in the case of the film projector[38] and various radio devices.

34 "Worries of Wartime," *The Wireless World* 3 (1915): 800.

35 "An Interesting Lecture"; "Military Aviation"; "Worries of Wartime."

36 Simeon, "Les Télécommunications et La Guerre de 1914 à 1918. Avènement de La T.S.F.," 40.

37 Peresypkin, *Radio – a Powerful Means of Defense of the Country [Radio – Moguchee Sredstvo Oborony Strany]*, 30.

38 Haidee Wasson, "Protocols of Portability," *Film History* 25, no. 1–2 (2013): 244, https://doi.org/10.2979/reseafrilite.47.2.09.

Moreover, mobile communication sparked a debate on other legal questions in the early 1910s that were only resolved after World War I. The British questioned whether a license was needed for railophone experiments and also what type of license was generally required for mobile communication.[39] Before the war, hardly any experts could agree on the future development of radio in the air and its role in international law. Several attempts were made at the 1906 and 1912 conferences about radiotelegraphy in Berlin and London, respectively, and at the Institute of International Law, where there was a specific project on wireless and aviation.[40] They never arrived at a consensus, however, for the technology was still evolving. Only after the war did they reach agreement, when the radio was already so advanced on airplanes that it required immediate regulation. With its spread into aviation, radio had significantly enlarged its sphere of influence from the sea to the air and had major implications for the notion of space.

Another important technological advance concerns the determining of location. World War I had demonstrated that radio technology could be used not only for intercepting a message but also for determining the location of the foe. This task was carried out with the help of radio direction finders, which determined the direction of the transmitting radio station. If the same radio transmission was received by several direction finders at different locations, one could determine the location of the transmitting radio station. Usually, the location of the radio station also indicated the location of the enemy's army.[41]

Thanks to such a direction finder, it was possible to facilitate navigation for military aviation. When radio stations for both transmission and reception had developed sufficiently to be installed on planes, they were used for navigation in the sky. Direction-finding sets were installed to intercept signals from airplanes so that they could determine their exact location. This location was then relayed to the airplanes, which could thereby find their way even at night. With the help of such direction finders, the Germans sent their Zeppelins on a raid to England. Similarly, the British also had modified direction-finding sets for airplanes, so they were prepared in advance for the attacks of the Zeppelins.[42] These direction-

---

39 BT Archives, "POST 30/2795C. Railophone System for Wireless Communication with Moving Trains (1910–1917)."

40 The Institute of International Law, "Project Drafted by the Institute of International Law for the Regulation of Aerostats and Wireless Telegraphy (1906)," *The American Journal of International Law* 7, no. 3, Supplement: Official documents (1913): 147–49.

41 Elizabeth Mary Bruton and Graeme Gooday, "Listening in Combat – Surveillance Technologies beyond the Visual in the First World War," *History and Technology* 32, no. 3 (2 July 2016): 213–26, https://doi.org/10.1080/07341512.2016.1231782.

42 Peresypkin, *Radio – a Powerful Means of Defense of the Country [Radio – Moguchee Sredstvo Oborony Strany]*, 31–32.

finding sets allowed for a better understanding of the position of armies behind enemy lines and were also an indispensable tool for intelligence services.[43]

Advances in military aviation equipment raised concerns about the development of voice transmission to avoid problems with piloting manoeuvres. Voice transmission was also helpful for the synchronization of troops, something that required a quick and straightforward method of communication but not necessarily for reaching far-away places. The invention of the Audion allowed for prompt and efficient communication by amplifying a radio signal; despite having had little success before the war, this innovation revealed its advantages throughout World War I. An influential figure in the spreading of this innovation was General Ferrié, who was in charge of the French army's military radiotelegraphic equipment. In October 1914, he assembled a team of engineers and physicists to work on the enhancement of the equipment. They were based in Lyons, which was further away from the war zone than the capital and also had several lighting lamp factories that could efficiently mass produce the new equipment. Already by the end of that same month, Ferrié's research team had designed the TM (*Télégraphie Militaire*) lamp, which had a similar design to the one devised by de Forest; it was also a triode (three electrodes) invention. In February 1915, the manufacturer Grammont began production on a small series, and by November 1918, production had reached 1,000 lamps per day.[44] The TM lamp was not a complete copy of the Audion, but it was an ingenious innovation boasting a highly stable and reliable performance, in contrast to the first Audion. Furthermore, the TM lamp allowed for a quick change of lamps thanks to its four-pin base, with all of these characteristics helping military radiotelegraphy to accelerate communication in combat, and to introduce voice transmission. Another important innovation was a transmitter that could produce a continuous wave. The first radiotelegraphic transmitters were called spark transmitters because they produced sparks; a shorter one was a dot, while a longer one was a dash; a continuous wave, on the other hand, was not intermittent and could carry sound. The first devices capable of producing a continuous wave were based on the invention of Danish engineer Valdemar Poulsen. In 1903, he had devised the arc converter, which was also called an arc transmitter or a Poulsen arc. It was more widespread in the US than in Europe; when it was introduced in Europe, almost all major governments and companies had already invested in other systems, those of Marconi and Telefunken among them, and only during the war were they compelled to invest more heavily in the continuous wave product.[45] Thanks to the

43 V.M. Cejtlin, *Intelligence Service Staff [Razvedyvatel'naja Sluzhba Shtabov]* (Cmolensk: Upravlenie voenno-uchebnyh zavedenij zapadnogo fronta, 1921).

44 Simeon, "Les Télécommunications et La Guerre de 1914 à 1918. Avènement de La T.S.F.," 25.

45 Marconi Archive, "MS Marconi 1798. HIS 150: Wireless Telephony – Early Tests to 1921" (1921).

war both of these products – the TM Lamp and the Poulsen arc – became important milestones in the development of radio, as investment in their production took place only out of military necessity. Production of these devices would be commercialized in the 1920s. After the war, the TM lamp became the most common piece of equipment for receiving radio, which also facilitated better broadcasting reception (Champeix 1980). The spread of continuous waves was also one of the necessary changes driven by the war because it greatly enhanced the quality of signals and allowed radio broadcasting to develop to its full capacity.

These three applications therefore show how radiotelegraphy came to transform the notions of space and time. The capacity of radio to spread across borders also created a new simultaneous experience in wartime, with Thompson's concept of 'despatialized simultaneity' (1995) best reflecting this capacity of radio communication to synchronize the moves of the belligerents.[46] This concept refers to the fact that telecommunication offered a unique experience of simultaneity that did not presuppose that people were at the same locality. The same events could be experienced at the same time but in different places, therefore uncoupling space and time. With new inventions and the development of military radiotelegraphy, the rapidity of transmission constantly improved while by the same token the notion of distance was also mitigated. Armies and fleets were able to act simultaneously and coherently even when separated by hundreds of kilometres. Moreover, this 'despatialized simultaneity' had a transnational character, as it moved across Europe along with the spread of the war.

The application of radiotelegraphy to airplanes led to the significant development of radio devices, in particular making them lightweight and portable and stimulating voice transmission. Those very characteristics would prove to be important, not least for the development of the radio broadcasting industry in the 1920s, and therefore could be interpreted as a path dependency of the new technology. Military necessity dictated that manufacturers apply themselves to the creation of light, small, and portable radio devices capable of voice and audio transmission, which became an important contribution to the development of radio broadcasting. The spread of voice and speech transmission was thus due to the military need to produce certain devices, but also to the quite complicated processes of transnational collaboration and inspiration, involving as they did inventors, factories, production, and distribution in a number of different countries.

---

46 J. B. Thompson, *The Media and Modernity: A Social Theory of the Media* (Stanford, CA: Polity Press, 1995).

Restrictions and transformations of the transnational networks

Destruction of radiotelegraph infrastructure

From the very beginning of World War I, radiotelegraphy revealed its importance and power for the belligerents. In an attempt to cut off all ill-intentioned agents, the governments had of course to restrict access to this mode of communication. To ensure control of the transmission of messages within a given country, European governments, one by one, issued orders requesting that private stations be dismantled for the war effort, with the main reason to avoid possible disturbances and to prevent messages of national interest from being intercepted by private individuals. In the UK, the Defence of the Realm Act, passed in August 1914, vested exclusive powers in the government to suppress published criticism, imprison without trial, and to commandeer economic resources;[47] amateurs in possession of wireless sets were being arrested as early as two days later.[48] Elsewhere, the Italian government suspended all licenses to set up and run wireless stations that had been issued to private citizens and companies in 1914 even before the country joined the conflict.[49] Some states did not pursue new laws due to existing legislation already in favour of state monopoly. For instance, in Russia, the legislation implied such strong bureaucratic restrictions on the installation of amateur stations that they almost precluded any such private activity, leaving radio communication de facto in the hands of the military authorities.[50] Usually, the restrictions on radio communication were enforced shortly after each country had entered the war. In the US, for instance, the ban was announced in April 1917 and completely aligned with the expectations and predictions of American radio amateurs.[51]

Radio was restricted not only by the warring parties but also in neutral countries, so as to prevent belligerents from gaining a particular advantage in radio communication by using stations located in neutral territory. The idea of neutrality was relatively young at the beginning of the twentieth century; it was discussed on

---

47 George Robert Jessop, *The Bright Sparks of Wireless* (Radio Society of Great Britain, 1990).

48 Norman F. Joly, *The Dawn of Amateur Radio in the U.K. and Greece* (London: Project Gutenberg, 1990).

49 Balbi and Natale, "The Double Birth of Wireless: Italian Radio Amateurs and the Interpretative Flexibility of New Media," 32.

50 Vladimir Ivanovich Shamshur, *The First Years of Soviet Radio Engineering and Radio Amateurs [Pervye Gody Sovetskoj Radiotehniki i Radioljubitel'stva]* (Moscow, Leningrad: Gosjenergoizdat, 1954).

51 "If We Are Closed Up," *Qst* April (1917): 26.

various occasions at international meetings and was informed by diverse international conventions and regulations about the neutral powers. Frequently, the concept of neutrality was synonymous with political independence; in fact, during the nineteenth century, these two terms were considered interchangeable, even in official documents.[52] Today, the idea of neutrality has lost much of its former importance and is considered a 'traditional law of neutrality'; after two World Wars, the systems of collective security and prohibition to resort to war were internationally accepted, therefore making a war internationally illegal.[53] However, at the beginning of the twentieth century, neutrality represented a crucial issue and encompassed radiotelegraphy as well.

The *Convention Respecting the Rights and Duties of Neutral Powers and Persons in War on Land*, signed in the Hague in 1907, imposed three main aspects of neutrality. First of all, the belligerents were forbidden to erect radiotelegraphic stations in neutral territories, as this could offer advantages to their forces on land or sea in the conflict.[54] This article was based on the experience of the Russo-Japanese War when the Russians had erected a radiotelegraph station in the seemingly neutral China, thus establishing radio communication with the government in the besieged garrison in Port Arthur.[55] The Convention also prohibited the military use of any stations established before the war in neutral territory. Finally, neutral powers were also required to look after the companies and private individuals in possession of telegraph, telephone, or radio devices. This article was also inserted in accordance with the Radiotelegraphic Convention of 1906 and, interestingly, the following declaration of London of 1909 contained yet more strict rules on maintaining neutrality in communication. However, it was not ratified, and countries entered World War I with the obligation to follow only these three points of the 1907 Convention: not to erect stations on neutral territories, not to use stations previously erected, and to control radio communication in private houses and offices.

Therefore, with the start of World War I, even the many countries that were not involved in the conflict were obliged to limit radio communication in order to maintain neutrality. Thus, both active participants in the war and neutral observ-

52 Balbi et al., *Network Neutrality: Switzerland's Role in the Genesis of the Telegraph Union, 1855–1875*, 32.
53 Diethrich Schindler and Jiri Toman, *The Laws of Armed Conflicts* (Leiden, Boston: Martinus Nihjoff Publisher, 2004); Kentaro Wani, *Neutrality in International Law: From the Sixteenth Century to 1945* (London and New York: Routledge, 2017).
54 ICRC, *Convention (V) Respecting the Rights and Duties of Neutral Powers and Persons in Case of War on Land. The Hague, 18 October 1907* (Hague: International Committee of the Red Cross, 1907).
55 Higgins, *The Hague Peace Conferences and Other International Conferences Concerning the Laws and Usages of War: Texts and Conventions with Commentaries*, 291.

ers were compelled to restrict radiotelegraph. A 'wireless silence' had swept across the globe:

> At first I thought that the bridge had been blown up, when suddenly I was astounded to see one of the wireless masts bend over and fall to the ground. I had scarcely recovered from my surprise when another explosion occurred more violent than the first, and a second mast fell. 'They are blowing up the station; the Germans must be near!' I said to myself.[56]

With these words, radio amateur Marcel G. de Gallaix shared his impressions of the dismantling of the station in Belgium, and so ended a station of transnational importance that had broadcast music and voice transmissions, as described in Chapter 2. Like many other European countries, the Belgian government was forced to destroy all its wireless stations, so as to prevent them from falling into the enemy's hands when Germany invaded; that station had been one of the most powerful in Belgium and was indubitably perceived to be a potential military asset. With the dismantling of the stations, many radiotelegraphy projects were abandoned. The Saturday Concerts in Laeken were interrupted as soon as war was declared.

In these attempts to destroy radiotelegraph stations, governments plainly recognized radiotelegraphy as a weapon. Headrick (1991) referred to radio as an 'invisible weapon,'[57] however, this weapon was in fact not as invisible as it seemed. The stations, with their bulky devices and high antennas, were not that easy to destroy. The tangibility and materiality of the seemingly 'invisible' infrastructure was problematic;[58] just as scholars have shown in the case of Internet cables, the transnational media infrastructure necessitates extraordinary efforts not only to be built, but to be dismantled later on. Moreover, even once destroyed, it leaves indelible traces on the fabric of urban and social life. Similarly, during the early stages of the war, numerous engineers and construction workers applied their unique skills to the task of damaging radiotelegraph constructions and equipment so that they could be of no further use by the enemy, with the process lengthy and involving several stages:

> It was only possible to carry away some of the light instruments; the remainder had to be destroyed. The most delicate parts were broken up with hammers, and to complete the destruction, the station was blown up with dynamite. The explosion was so violent that part of the granite parapet was broken, and a large crack opened in the roof of the tunnel. Finally, so that even the ruins could not be put to any possible use, the station was filled with straw

56 Marcel G. de Gallaix, "The Brussels Station. Wireless Telegraphy in War," *The Wireless World* (February 1915): 717, https://doi.org/10.2307/1323303.

57 Headrick, *The Invisible Weapon: Telecommunications and International Politics, 1851–1945.*

58 Starosielski and Parks, *Signal Traffic.*

and hay and set on fire. A dense smoke rose from the tunnel. It was seen rising over the canal until the evening, and the last bursts of flame were not extinguished when a detachment of the enemy's cavalry appeared on the scene.[59]

All around Europe, radio antennas were destroyed. Without a doubt, the destruction of such massive constructions was disastrous for the many innovative projects that had just started or were about to be launched. In particular, it had cost many inventors a pretty penny and much effort to establish these radio stations. The demolition of the Brussels station, for example, destroyed the investment, sufficient as it was to see the project through to completion, and more than three years of patient research by engineer Robert Goldschmit.[60]

Many innovative engineers suffered from the restrictions brought by the war. Requests to dismantle stations were indeed fatal to some inventors, many of whom would never resume their projects after the war. One of the most representative examples is the Galletti Wireless Telegraph and Telephone Company, which suffered losses not only in its home country of Britain but also in France. Galletti was known as an inventor of an exceptionally efficient radiotelegraph system. In 1911, he had constructed a transatlantic wireless communication service for the Indo-European Telegraph Company, with his success and the requests of other companies to construct similar systems leading him to establish the Galletti Wireless Telegraph and Telephone Company, which had an office in London. The central location of his transatlantic transmitter was, however, in Champagneux in the Rhône-Alpes region in France, due to its geographical and geological particularities. However, the success of his company was unfortunately cut short. Soon after the transmissions from Champagneux started, the war broke out, and the French Government shut down the station. Galletti's equipment was dismantled and stored in Bordeaux, where it was not put to any practical use; after it was released from custody, it was so damaged that it was no longer viable.[61] The same fate befell other radiotelegraph stations in Britain. The Devizes, Chalvey and Slough wireless stations were taken over by the War Office and Admiralty; after the war, Galletti even claimed compensation for them,[62] however, he was never able to attain the same commercial success, on account of the damage done to the equipment and a change in the socio-political context. His company eventually shut down in November 1925.

59 Gallaix, "The Brussels Station. Wireless Telegraphy in War," 717.
60 Ibid., 718.
61 Ken Beauchamp, *History of Telegraphy* (London: The Institution of Engineering and Technology, 2008), 198–200.
62 BT Archives, *POST 30/2425A. Experiments and Licences for Galletti Wireless Telegraph System, Part 2. 1914–1919* (London, 1914).

Thus, at the start of the war, and at the political level, numerous countries (both warring and neutral) took measures to limit communication by radiotelegraphy. Indeed, radiotelegraphy became one of the many prerogatives of the military and secret services, which was also reflected in the popular songs of the time. One of the well-known songs of World War I was 'Send me a Line when I'm across the ocean,' which dates from 1917, was published in the US, and distributed in the UK, Australia, and New Zealand. The lyrics describe an American soldier leaving for war who requests his lover to write him 'a letter nice and long' to express love and tell of the things that would be happening without him around. There was nothing about wireless or sending a kiss via wireless but only a request for a traditional and reliable communication to be sent by post from 'somewhere in France,' as the cover suggests.[63]

Transnational projects paused or cancelled

Many transnational projects suffered from wartime restrictions. In particular, the most symbolic places for transmitting transnational information were occupied by the armies. The French Army used all the French radiotelegraphic stations, including the most famous one – the Eiffel Tower. The same fate befell most of the stations in Europe that had previously been included in the transnational networks of time signals and weather reports. Some stations continued their regular transmissions, but these were not regular enough, and they also did not have as many listeners as before. In Switzerland, for instance, clockmakers complained so much about their inability to use radiotelegraph time signals that the Swiss government had to employ a wired telegraph and telephone to transmit time signals for clockmakers during the war. In this case, the Swiss state assumed responsibility for controlling transnational communication and restricted the opportunity for international time signals to be sent from the Eiffel Tower and other radio stations, replacing them with a new and limited national service.[64]

The war had also damaged meteorological services. Even a minor disruption in this coherent network affected the speed at which a message was received, which could greatly slow down the transmission; if the information did not make it in time to be included in the meteorological report, the transmission of the message was not of any use.[65] The weather forecasts had long since been stopped, but

63 Irving Crocker and George L. Cobb, *Send Me a Line When I'm across the Ocean [Song]* (New Yor: Walter Jacobs, 1917).

64 "Telephonic Time Service in Switzerland," *The Wireless Age* March, 5, no. 6 (1918): 488.

65 "Wireless Weather Report," *The Wireless World* (November 1914): 522.

they would resume in some countries in 1917, New Zealand among them.[66] However, the overall disruption of a coherent network did not allow radiotelegraphy to be used to its full potential; as a result, the newly introduced network of stations spreading time signals and meteorological reports also failed. Some of its parts were still functioning, but the stations were never fully open to the newest technological developments. The coherent and transnational network had been massively disrupted, and transnational projects lost their potential to grow.

During World War I, news services became an essential instrument for the dissemination, receipt, and exchange of information, especially when cables had been destroyed. The most powerful and best known radio stations, and the distributed network of news correspondence in many European countries, allowed the Wireless Press, operated by the Marconi Company, to offer a particular service, War News, which became the most important news source for many different newspapers and press agencies.[67] The Wireless Press war news consisted of a collection of the messages received by radiotelegraph from foreign countries, including allied, neutral, and enemy ones, and then translated. Newspapers and publishers would pay a fee to subscribe to this service and receive news reports in printed form; among them were the *Daily Telegraph, Evening News, Glasgow Herald, Birmingham Daily Mail, the Times, the Sunday Chronicle*, and various others. The press agencies, such as the New York American, the United Press Association, and the Associated Press, also used this source of information to relay the news to their clients.[68]

These news agencies were mostly under governmental control, even though they were usually private companies. All messages coming from the Wireless Press were filtered and censored; before sending information to newspapers and press agencies, the Wireless Press had them checked by the British Admiralty, with every message bearing the caption, 'British Admiralty intercepts collected by the Wireless Press.' Furthermore, in the newspapers, these messages had to be published as written and be preceded by the above description.[69] It was the Wireless Press service that had received the breaking news about the Russian Revolution in 1917 and passed it on to the government. The fragmented pieces of news that reached England over the radio waves were not clear enough and required additional interpretation, so the news was not passed immediately to the newspapers but instead to officials only. Therefore, for a long time, the Wireless Press

66 "New Zealand," *The Wireless World* (February 1917): 69.

67 "List of Subscribers to War News Sept: 1915" (University of Oxford, Bodleian Library, ms. Marconi 301, 1915).

68 Ibid.

69 Marconi Archive, "Serial C. 3021. Press Bureau," December 9, 1916.

played the role of a mediator, sending information from radio telegrams to news agencies.

Throughout the war, this news service became such a relevant and useful communication tool that in 1916 the Wireless Press suggested to the newspapers that it organize the radiotelegraph service directly: to 'radiate propaganda by wireless' because it represented the only means of communication that 'secured simultaneous publicity over a very wide area.' Furthermore, the statement underscored that the newspapers themselves had formulated this idea.[70]

The notion of organizing such propaganda messages was, however, decidedly transnational. Transnational communication was still of greater importance than the task of informing the national public; the primary goal of the Wireless Press was in fact to distribute the foreign news, not the national news, and it drew up contracts with powerful radio stations in Italy, Russia, Romania, and Norway to ensure sufficient news distribution. The authors of the suggestions underscored the following: 'By tapping a key set at the Carnarvon Station, messages can be received simultaneously at any point in Europe, in Egypt or on the American coast, and these stations can relay on to other stations and so cover the greater portion of the civilized world.'[71]

Using Benedict Anderson's term, the 'imagined community' of the radiotelegraph was not a nation but instead the whole world,[72] however, this idea could not be realized in its full force, and this for many reasons. First of all, the war imposed certain limitations on the use of radiotelegraphy. Second, the realization of such a project would have put the Marconi Company in an advantageous position on the market, as it was the only business that could afford to invest in such an extensive and strong network, with such a prospect not widely welcomed. Another difficulty was the timing of transmissions; the messages would indeed reach the newspapers quite rapidly, but this temporal advantage of wireless was negated by the fact that the newspapers were not so fast, their editions always being issued in the morning or the evening.

70 "The Wireless Press Limited Established to Carry on the Business of Publishers . . ." (University of Oxford, Bodleian Library, ms. Marconi 301, 1916).

71 Ibid.

72 Benedict Anderson, *Imagined Communities: Reflections on the Origin and Spread of Nationalism* (London and New York: Verso, 1983).

International conferences deferred

For the duration of World War I, international collaborations were paused, and many international conferences were deferred or even cancelled. The International Conference of Engineers of Telegraph and Telephone (Congrès international des ingénieurs des Telegraph et des Téléphones) had been scheduled for Bern in September 1914 but was postponed to a 'more favourable time,' which was mistakenly believed to be the spring of 1915.[73] As infrastructure conditions were immediately interrupted after the war's outbreak, some 195 copies of the papers prepared by the British Delegation for this meeting in Bern did not reach their destination due to being stopped at Ghent and then returned to London.[74] The international telegraphic conference, which was planned for Paris in 1915, was also postponed because of the French government's decision.[75] The International Radiotelegraph Conference scheduled to take place in Washington in 1917 likewise never took place.

It should be noted that war was always a possible scenario in the minds of people living at that time. In 1903, international discussions about radiotelegraphy were already focused on the duties of neutral countries towards belligerents, and vice versa.[76] A special article 21 of the 1906 and 1912 ITU conventions stated that only the stations of public correspondence should follow international regulations, while those of naval and military communications could be used according to the governmental interest if they did not intrude upon the communication of others and continued receiving distress calls.[77]

Therefore, with the closure of public stations and the rise of military ones, the ITU regulations were becoming irrelevant for an increasing number of stations when they entered the war which is why, on a legal level, the ITU had to distance itself from the war arena. The communication and networking around the ITU from 1914–1915 was predominantly focused on gathering the most recent information about the radiotelegraph. Countries asked about national achievements in radio, such as the lists of stations and statistics, however, the ITU Bureau was not always able to provide relevant information about the current state of

73 "Revue Télégraphique de 1914," *Journal Télégraphique* 1 (1915): 4.

74 BT Archives, "POST 30/2921B International Telegraph Union Informal Conferences of Technical Experts," 1914.

75 BT Archives, "POST 30/3089A International Telegraph Conference, Paris in 1915, Part 2," 1915.

76 ITU Archives, *Documents of the Preliminary Conference on Wireless Telegraphy (Berlin, 1903)* (London: George Tucker, 1903), 41.

77 ITU Archives, "Convention Radiotélégraphique Internationale (1906)"; ITU Archives, *International Radiotelegraph Convention (London, 1912).*

affairs in radiotelegraphy, as the belligerents in the state of war were not sharing information. The statistics of 1914, for example, did not include the data about the number of radio stations in most European countries, such as Germany, Great Britain, and Italy.[78]

The official monthly periodical of the ITU, *Journal Télégraphique*, aimed to be neutral and distant from political conflict. It was suspended from July 1914 for five months and reinstated in January 1915, with only oblique references made to the war, in terms of 'the political events in Europe,' and no specific materials about the war included in the 1915 issues. In this international arena, the war seemed not to be a legitimate topic. Only a brief note in the issue of January 1916 offered a fleeting glimpse of the war. Editors wrote: 'Last year, the war continued in Europe. This time is not likely to stimulate the development of international relations. Therefore, it is not surprising that in this review one cannot find significant news related to telegraphy.'[79]

During the war, the ITU experienced a shortage of information because some of its staff were mobilized. With the war in full swing, on the first page of the January 1916 issue, the editors explained: 'It is unfortunate to note, moreover, that the numerous service restrictions introduced in 1914 to international correspondence, as a result of the war, were maintained in 1915, and that several new restrictions were added to it.'[80]

Many projects at the ITU were deferred. For instance, the establishment of a memorial was scheduled for May 1915 but was not realized until December 1922. Furthermore, during this time, the prices of the materials and work had increased appreciably, and therefore the total cost of the monument had exceeded the reserved budget.[81]

During the war, the ITU was in an ambiguous position. Even though circumstances did not favor international collaborations, the international organization never interrupted its activity. The Bureau sustained its networks, communicated information, and functioned overall. From 1914–1915, the *Journal Télégraphique* preferred to publish information about inventions from previous years, which it had to hand. The information coming from the ITU was often issued roughly two years after the actual events, as this much time was needed to compile and analyze these different national data, which is why for 1914 and 1915, the ITU could use previously collected materials and explain the absence of current news in ra-

78 ITU Archives, *Statistique Générale de La Radiotélégraphie Dressée d'après Des Documents Official. Année 1914* (Bern: Bureau International de l'Union télégraphique, 1916).

79 "Revue Télégraphique de 1915," 3.

80 Ibid.

81 ITU Archives, "Circular No. 752," 1923.

diotelegraphic development by calling them 'mediocre.'[82] The journal reprinted some recent news from other print media, as the ITU did not receive any official notices.

The following chart demonstrates the amount of the outgoing and incoming correspondence of the ITU (Figure 18). It reveals that the two biggest peaks in communication with the ITU frame the war, signifying the rise of international collaborations: the 1912 conference provoked the first; the second, in 1920, denotes attempts to deal with the consequences of the war. During the war, on the other hand, ITU sustained normal levels of communication activity, supporting its network.

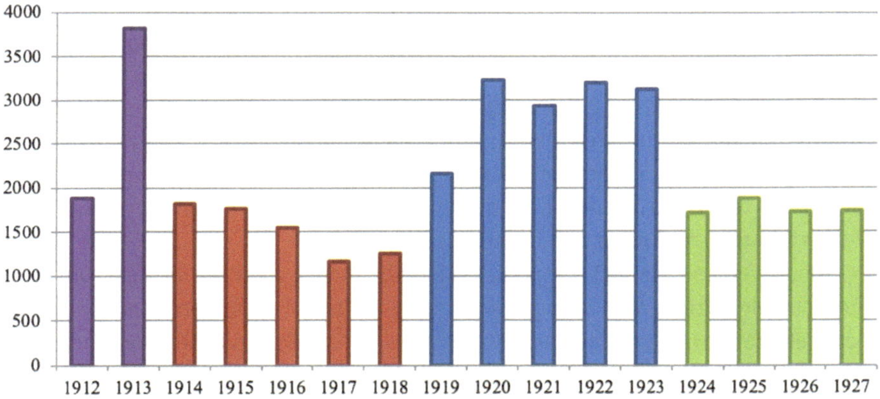

Figure 18: The amount of ITU correspondence about the radiotelegraph, 1911–1927. Based on data from the ITU Archives *Correspondance: Radiotélégraphie*, 1912–1927.

The ITU was functioning, responding to letters, and providing regular publications throughout the entire war. What was that communication about, if many international projects on wireless were disrupted by the war? With the shortage of materials about European wireless, ITU had to refocus the materials. Functioning as an 'antenna' for news about radiotelegraphy, the ITU included in its agenda more materials from what was understood to be an 'extra-European' space. The need to draw attention to developments beyond Europe revealed to a wide audience the fact that there were important inventions outside European politics, where the radiotelegraph had spread as a truly transnational technology. An increasing number of countries engaged in international collaboration with the

82 "Revue Télégraphique de 1915," 1.

ITU, with the adherence of many countries to the 1912 Regulations which occurred after the conference as follows: Mexico (in 1913), Panama (1914), Guatemala (1914), Colombia (1914), Bolivia (1915), Peru (1915), Cuba (1918), Iceland (1919), Venezuela (1920), Ecuador (1920), China (1920), and many colonies that had not appeared originally in the Union as participating countries.[83]

The colonies themselves also had their own interests regarding wireless development. Even though the empires typically overrode them, radio development in colonies often followed its own path despite European politics, with the colonies writing to the ITU separately and presenting their information independently. These territories were not only dependent on Europe but also provided valuable material which was of real interests to experts elsewhere. During World War I, the ITU had drawn attention to these territories, indicating that they were not only followers of imperial inventions but also had regional or even local characteristics where technological development was concerned. Australia, for instance, was troubled by the monopoly of the Marconi Company in the British Empire and refused to have anything to do with its business.[84] Australians questioned whether having the radiotelegraph as a governmental monopoly was indeed the most effective guarantee of the speed, flexibility, accuracy, and secrecy of the radiotelegraphic service in their country.[85]

Moreover, commercial companies gradually moved their businesses from risky and unstable Europe to more neutral territories. Thus, Telefunken was able to protect itself from tremendous losses by transferring its reserves to Holland and developing its subsidiaries in Argentina, whose President, Victorino de la Plaza, had declared neutrality. Throughout the war, radiotelegraphy developed rapidly there, and in 1916 Argentina established a stable communication between Buenos Aires and New York City.[86] In 1918, the radiotelegraph station set up by the Minister of the Navy at San Julian was opened to public service, which substantially reduced the duration of the transmission of messages between the extreme south of the continent and Buenos Aires. With the opening of radio stations at Punta Delgada, the coast of Patagonia was also put into radiotelegraphic communication.[87]

During the war, radio stations in this extra-European space were seen in relation to their connection to the global radiotelegraphic network. For instance, an account of the newly established radiotelegraphic station in Tahiti in February 1916

---

83 "L'Union Radiotélégraphique," *Journal Télégraphique* juillet (1922): 124–26.
84 John Adair, "The Marconi Contract," *The Commentator*, January 22, 1913.
85 Marconi Archive, "Government -v- Private Control of Imperial Wireless Stations" (University of Oxford, Bodleian Library, ms. Marconi 301, 1916).
86 "Nouvelles Installations de Télégraphie sans Fil," *Journal Télégraphique* 12 (1916): 262.
87 "Argentine," *Journal Télégraphique* 1 (1918): 16.

remarked upon its capacity to communicate with different parts of the world: California, Australia, South America and Asia.[88] In September 1916, the successful linking of stations in Funabashi (Japan) and San Francisco was hailed, as an instance of transpacific communication;[89] as transatlantic communication was already in use, this connection meant that radiotelegraphy was now able to span the entire globe. Sometimes, the ITU also published materials that it had not been officially notified about. Thus, it reported that radiotelegraphic communications had allegedly been established between America and Norway, America and Sweden, Russia and Japan, and Japan and Hawaii. It also reported that the international radiotelegraphic network would soon link up with China, as radio stations were operating in Kalgan, Beijing, Woosung, and Guangzhou, while another was under construction in Wuchang. Underscoring the fact that the ITU had not been officially notified about the erection of these stations, it also informed its readers that the editorial board would reconfirm this information 'after the restoration of the peace.' The war thus did not only lead to a dearth of content but also allowed the ITU to hint or insinuate.[90]

While European countries were engaged with the war and concentrated on obscuring any advances in wireless technology they may themselves have made, to their own military advantage, other states were prepared to reveal more to the world regarding how wireless was developing. The presence of non-European actors became more prominent, and inevitably had an impact on the agenda of transnational wireless; from having an essentially European character, it shifted, or rather expanded, to being a truly global technology. The proportion of the communication of ITU with non-European countries markedly increased during the war,[91] and consequently, the International Radiotelegraph Union materials, such as recent news, became less Eurocentric. From a global perspective, this shift meant that different social groups were receiving this global vision of radiotelegraphy. The recipients of the news from the ITU were not limited to their list of subscribers; indeed, such news was spread by numerous technical and engineering associations and societies, such as the Société Française Radioélectrique in

88 "La Station Radiotélégraphique Côtière de Tahiti (Océanie)," *Journal Télégraphique* 2 (1916): 47.

89 "Radiotélégraphie Transpacifique," *Journal Télégraphique* 9 (1916): 216.

90 "Revue Télégraphique de 1915."

91 See more on IRU communication during the war in Maria Rikitianskaia, "The International Radiotelegraph Union Facing World War I, 1912–1927," in *History of the International Telecommunication Union (ITU). Transnational Techno-Diplomacy from the Telegraph to the Internet*, ed. Gabriele Balbi and Andreas Fickers (Berlin: De Gruyter, 2020), 191–214.

Paris[92] and the Russian Society of Telegraphy and Radiotelegraphy in Saint-Petersburg.[93] On many occasions, national publishers would themselves republish the ITU circular letters or news, and quite often the information was passed on in turn by many individuals or institutions, filtering down to local radio amateurs' groups and local inhabitants.

This shift signifies the change in the networks of actors around the ITU, which were directly influencing the understanding of radiotelegraphy on a global level. In the STS terms, this change illustrates the transformation of the relevant social groups over the course of the war, which inevitably served to alter the general understanding of what wireless was. This observation about a different dynamic in European and non-European countries also aligns with the recent 'global turn' in World War I studies. As recent historical studies have shown, in Europe, World War I was also determined by the non-European sites of battles, namely those that involved millions of non-white men mobilized both for combatant and non-combatant roles.[94] Scholars are now agreed that World War I affected finance and trade globally.[95] The war also brought into view different kinds of racial, national, religious, and class identities from every corner of the globe. World War I was indeed a world war, even if it had a Eurocentric character.

Similarly, the period of 'wireless silence' in the European arena nevertheless led to discoveries; there was an active development of radio beyond Europe as well, which began with European initiatives but soon found its own way. Communication with the colonies and independent countries far from Europe thus intensified, demonstrating that the extra-European environment could be a space where international collaboration happened and was made possible by radio stations connected to a global radiotelegraphic network.

92 ITU Archives, "D. 16. N. 1. Société Française Radioélectrique," in *Registres de Correspondance: Radiotélégraphie* (1911); ITU Archives, "D.8. N. 33. Société Française Radioélectrique," in *Registres de Correspondance: Radiotélégraphie* (1919).

93 ITU Archives, "D. 19. N. 79. Soc. Russe. de Télégr et Téléph. sans Fils," in *Registres de Correspondance: Radiotélégraphie* (1914); ITU Archives, "D. 19. N. 15. Soc. Russe. de Télégr et Téléph. sans Fils," in *Registres de Correspondance: Radiotélégraphie* (ITU Library & Archives, 1917).

94 Santanu Das, "Introduction," in *Race, Empire and First World War Writing*, ed. Santanu Das (Cambridge, New York: Cambridge University Press, 2011).

95 Hew Strachan, "The First World War as a Global War," *First World War Studies* 1, no. 1 (2010): 3–14, https://doi.org/10.1080/19475021003621036.

Amateurs beyond the battlefield: Thousands of 'listening ears'

In 1914, 21-year-old British amateur W. Kenneth Alford overheard a radio message, which stated: 'Aug. 4 1914. POZ (Nauen). War is declared against France and Russia.' The news was striking, but the irony was that neither his friends nor neighbours believed it until they read the newspapers the following day. In 1918, however, they did not for a moment doubt the radio message recorded in W. Kenneth Alford's log-book as, 'Nov 11 1918. FL (Paris). War is over'. The evidence for this story is uncertain, however, it has been reprinted in many reflections on the role of radio amateurs over the course of World War I.[96] This kind of anecdotal storytelling is representative of narratives about the development of a new medium, which Natale[97] calls the 'biography of media.' In particular, anecdotes of this sort frequently serve to help people make sense of new experiences and events and cope with changes in media. This specific anecdote is a representative example of the genre, as it shows a previously unknown method of radio communication becoming a reliable source of information over the course of the war.

There were so many amateurs and experts who had benefited from training and gaining access to the new devices that the time came when the army no longer needed many technically knowledgeable amateurs. Hobbyists showed their dissatisfaction at this in their letters: 'I naturally thought that my knowledge would be of some little use in some sphere or other at the present time, and made several enquiries, but only to receive the 'cold shoulder' at every attempt.'[98]

This governmental monopoly and its restrictions hit radio amateurs particularly hard, leaving them without a hobby to practice. Radio amateurs quickly fell into disgrace for two reasons; first, they had the ability to intercept secret messages and second, even a small and unintended mistake in the construction of a radio receiver could turn it into a transmitter and would be of benefit to a spy. In some countries, like the UK, some radio equipment was confiscated, only to be returned to the amateurs when they re-applied for a license after the war.

A 'wireless silence' was supposed to begin, as radiotelegraphy was now the exclusive preserve of the military. However, radiotelegraphy was not wholly abandoned, the situation in fact becoming somewhat more nuanced and complex. The disruption of the infrastructure and, more importantly, the disruption of the trans-

---

96 See, for example, "Farewell," *Electronics & Wireless World* 92 (1986): 60.

97 "Unveiling the Biographies of Media: On the Role of Narratives, Anecdotes, and Storytelling in the Construction of New Media's Histories," *Communication Theory* 26, no. 4 (2016): 431–49, https://doi.org/10.1111/comt.12099.

98 J.R. Lewis, "Correspondence," *The Wireless World* (July 1916): 215.

national infrastructure, did indeed bring changes to media business models, however, these changes never entirely suppressed the growth of radio among civilians.

Despite the restrictions and the full mobilization in Europe, there was still amateur activity beyond the battlefield. First, even though the number of print publications had been substantially reduced, some literature continued to be available to amateurs. A considerable number of journalists and editors had responded to the call of duty, but many journals were still being published.[99] Monthly magazines did, however, become considerably slimmer because of the shortage of printing paper and materials; for instance, *The Wireless World* in Britain ran to only 58 pages instead of its usual 72. Some periodicals were unexpectedly shut down. In September 1914, the Russian monthly *Herald of Wireless Telegraph* apologized for the deferral of publications for more than three months and expressed hopes for an imminent resumption and return to regular production; unfortunately, after that forward-looking statement, the journal was permanently suspended. Later, other new magazines appeared, such as *Vestnik voennoj telegrafii* (Herald of military telegraph), but they already had a new focus: military usage and the war's impact on radiotelegraphy.[100] Even the monthly ITU *Journal Télégraphique* became slimmer and was printed on thinner, cheaper paper, but it continued to be issued and sent to amateurs and experts around the world.

Overall, the restrictions in place did not lead to radio disappearing from everyday life; indeed, society became still more fascinated by it. For example, instructions for receiving signals from the Eiffel Tower were published in 1915 in the midst of the war and even included a dictionary useful for an amateur unacquainted with the French language.[101] There were still amateurs who followed the news about recent inventions and were quite active in this field, although officially their activity was more theoretical than practical. Without the possibility of establishing contact by wireless, the members of wireless societies were afraid of losing touch with each other; in some societies, meetings became even more frequent than before.[102]

Furthermore, in all prohibitions and restrictions, there were still small loopholes that allowed amateurs to operate during the war and improve their techniques. In the UK, the telegram about the suspension was sent only to those who

99 "The Effects of the War," *The Wireless World* (September 1914): 355.
100 "From the Editorial Board [Ot Redakcii]," *Herald of Military Radio Telegraphy and Electrical Engineering [Vestnik Voennoj Radiotelegrafii i Jelektrotehniki]* 1 (1917): 1–2.
101 Bureau des longitudes, *Wireless Time Signals. Radio-Telegraphic Time and Weather Signals Transmitted from the Eiffel Tower, and Their Reception [Authorised Translation].*
102 "Questions and Answers," *The Wireless World* (September 1914): 409–10.

used more than 50 watts of power;[103] subsequently, some amateurs remained in possession of their transmitting and receiving equipment. Even with the imposed restrictions, radio amateurs in various European countries continued operating. In Britain, for instance, private station owners were informed of the decision to dismantle all experimental wireless stations by the following communication: 'Dear Sir. – In accordance with your wireless license, the Postmaster-General requires you to remove at once your aerial wire and dismantle your apparatus. One of his officers will call upon you.'[104] However, if no one did call upon the operator, an amateur might here and there still be in possession of his precious equipment. Moreover, some constructed new stations.

There were also amateurs who moved to other countries and continued operating and kept in contact with their native amateur community. For instance, one Englishman installed a ham radio station in Turkey and shared his experience of listening to messages in 1915 with his pals on the pages of *The Wireless World*; he described listening to the foreign Embassies and their vessels, predominantly in German and French, with their communication sounding to him like a 'concert of Europe.'[105] This 'concert of Europe' was heard by hundreds of hobbyists in different European countries. In general, amateurs were mainly picking up and listening in because they wished to remain undetected,[106] and most amateur radio sets were unable to transmit.[107] The amateurs described as 'ether-flâneur' this practice of listening when the broadcasts were not regular, were chaotic, and were not addressed to them. The closest modern-day approximation to this practice is the modern notion of channel surfing, the time-consuming and uncontrolled listening to a medium, such as TV zapping.[108]

Some amateurs also took the risk of operating illegally. The journals that managed to remain in publication throughout the war asked the amateur communities to desist from any illegal activity, however, court records nonetheless show how extensive the activity of amateurs in many countries was;[109] *The Wireless World*

---

103 Clarricoats, *World at Their Fingertips*, 36.

104 "Wireless in the War," 381.

105 The Near East, "Wireless Amateurs in England Are . . .," *The Wireless World*, no. 3–33 (December 1915): 595.

106 Bruton and Gooday, "Listening in Combat – Surveillance Technologies beyond the Visual in the First World War."

107 Balbi and Natale, "The Double Birth of Wireless: Italian Radio Amateurs and the Interpretative Flexibility of New Media."

108 "Interaction versus Consumption: Mass Media and Art from 1920 to Today," in *Timeshift: The World in Twenty-Five Years* (Ostfildern-Ruit: Hatje Cantz, 2004).

109 Balbi and Natale, "The Double Birth of Wireless: Italian Radio Amateurs and the Interpretative Flexibility of New Media."

even launched a special section called 'Wireless in the Courts' in 1915 and 1916.[110] Many amateurs were convicted because of a suspicion of espionage; for example, Sergey Zhidkovsky, a radio amateur from the southern part of Russia (currently Ukraine), was arrested in 1914 because he regularly received international broadcasts from the Eiffel Tower, transmitted weather reports by wire to Kiev, and communicated with foreign countries in general.[111] Some amateurs continued operating discreetly and tended not to leave any traces of their activity. In the early days of broadcasting, radio experimentation entailed transgression of a sort 'that would throw governmental sovereignty into question';[112] the main activity was to listen in and secretly improve the technology and devices, which also explains the lack of information about the activity of amateurs in the literature.

Sometimes illegal operations of this sort were not a deliberate form of resistance but rather just how things were. Before the war, not all amateurs were licensed, which complicated the process of finding them in many European countries. In the UK, a pioneering country in organizing hobbyists, only 69 out of 162 members of the Wireless Society of London held transmitting licenses, which means that at least half of these members knew how to operate the devices and conducted experimental work without a license.[113] However, in the 1920s, many private individuals and institutions were broadcasting without a permit because their countries had not established a licensing system.[114]

Overall, the radio amateur community showed great resistance to the restrictions and limitations and continued to operate and publish magazines. Amateurs desperately tried to preserve their stations, with one man reporting to have a wireless station naively explaining that of the wires alluded to, one was a clothes line and another a perch for birds.[115] Some amateurs still secretly experimented on their home stations because it was already possible to receive wireless signals utilizing indoor aerials (though the signal was much worse than that obtained from an outdoor aerial); these amateurs were explicitly addressed in a statement of Home Secretary Mr MacKenna, which announced that the Post Office had established a special system for wireless detection.[116] In March 1915, *The Wireless*

---

110 "Wireless in the Courts," *The Wireless World* 3, no. 32 (1915): 523.

111 Jurij Dobrjakov, "The Case of the Peasant Zhidkovsky [Delo Krest'janina Zhidkovskogo]," *Radiofront* 13 (1935): 7–8.

112 Daniel Gilfillan, *Pieces of Sound. German Experimental Radio* (Minneapolis, London: University of Minnesota Press, 2009), xvi.

113 Elaine Richards, *Radio Society of Great Britain Centenary. 100 Years Working for Amateur Radio* (Bedford: Radio Society of Great Britain, 2013), 4.

114 Lommers, *Europe – on Air: Interwar Projects for Radio Broadcasting*, 165.

115 "War Notes," *The Wireless World* (November 1914): 505.

116 "Wireless Police," *The Wireless World* (November 1914): 519–20.

World noted that the journal had received numerous letters from radio amateurs who were still conducting experiments, although this activity was prohibited without the written permission of the Postmaster-General. The editors specifically called upon these readers to stop experimenting, with this call repeated later in June, which indicates that not everyone had obeyed the law. Some of them excused their hobby as the development of new techniques that would be useful for the country's victory in case the military happened to use the private station.

In 1916, some amateur activity showed signs of stagnation, with various contributory factors. Some radio amateurs were called up while others had their equipment confiscated, while others began attending engineering schools and other organizations and shared their apparatuses. Clubs and societies still held regular meetings, and some even obtained permission to work on an apparatus without making actual communications and always under controlled conditions. In the absence of real practice, however, theoretical work was not so appealing. In August 1915, *The Wireless World* encouraged readers to proceed with theoretical work while waiting for the end of restrictions.[117] To sustain their practice with Morse Code, amateurs were advised to construct a busser set for intercommunication between two operators sitting in separate rooms of the same house.[118] The Irish School of Wireless still offered training and held amateur gatherings using an apparatus that could not receive or send messages and was therefore without actual radio communication. Moreover, the Wireless Press, Ltd. and the Gramophone Co., Ltd. produced a series of records that could be reproduced on any disc-talking machine using the ordinary needle method of reproduction and which gave signals in Morse characters, thus allowing amateurs to practice first-class Morse sending at various speeds without breaching the Defence of the Realm Regulations.[119]

American amateurs were better placed in this regard than their European counterparts. After the war started, President Wilson immediately proclaimed America's neutrality; according to the Hague Convention of 1907, which America had signed, the neutral power was called upon to take any measures to control all radio companies and private individuals. On August 5, President Wilson issued an order: 'All radio stations within the jurisdiction of the United States of America are hereby prohibited from transmitting or receiving for delivery messages of an

117 "Amateur Work during the War," *The Wireless World* (August 1915): 322–23.
118 C.C. Barnard, "A Morse Practice Set," *The Wireless World* (November 1914): 528.
119 "Wireless Signals for the Home," *The Wireless World* (March 1916): 794–97.

unneutral nature, and from in any way rendering to any one of the belligerents any unneutral service, during the continuance of hostilities.'[120]

Therefore, when the US still did not enter the war, the amateurs experienced very few restrictions. In contrast to European amateurs, Americans were not massively mobilized, the radio stations were not immediately dismantled, and the networks continued functioning. In fact, the war spreading across Europe caused American amateurs to mobilize their resources to come up with a plan to help the country. In November 1915, a group of amateurs announced the creation of the Radio League of America, with one of the aims of this organization to aid the armed forces in case of attack. In an address to the radio amateur community, the founders of the Radio League of America underlined the advantages of the radio among other media technologies:

> There might not be a telegraph or telephone line around for miles, or if it did exist, it is certain that spies operating on land would have found little trouble in putting it out of commission beforehand. However, there will be a lone radio amateur on the alert who has seen the approaching fleet and within 30 seconds Washington will have the priceless intelligence.[121]

Furthermore, American editors valued amateur radio the most, and reflected on Europe in the following way: 'If France or Belgium had possessed an effective amateur wireless scout service there might possibly be a different story to tell to-day.'[122]

During this period of US neutrality, amateurs helped the Navy and made a significant contribution. Mr. Charles P. Apgar, an amateur wireless experimenter, recorded the signals of the Sayville Station operated by the German Embassy and proved that the station had transmitted coded and not neutral messages to Germany. An operator taking down the message by ear could easily miss the code because it was hidden in the order of repeating each word to make the symbol clear, but Apgar possessed a loudspeaker and a phonograph, which he used to record the voice for two weeks. At that time, few believed in phonograph recording; even the Secretary of the Atlantic Communication Company, Dr. Frank, commented on the news: 'that Mr. Apgar can record messages sent out by wireless on

120 Woodrow Wilson, Gerhard Peters, and John T. Woolley, "Executive Order 2011—To Enforce Neutrality of Wireless Stations," *The American Presidency Project*, August 5, 1914, http://www.presidency.ucsb.edu/ws/?pid=75364.

121 H Gernshack, "The Radio League of America," *The Electrical Experimenter* December (1915): 382.

122 Ibid.

a phonograph cylinder is hardly worth discussing. That is physically impossible. I have never heard of its being done.'[123]

The American amateur movement expanded massively up until 1917. In January 1917, the *QST* journal mentioned that in no previous year had amateur wireless progressed as much as it had in 1916. There was not only an increased number of amateur stations but the quality of transmission had itself been enhanced; there were already several hundred stations capable of reaching a distance close to 1,000 miles instead of just 500. There was even regular broadcasting, such as the evening experimental Highbridge Station of Lee de Forest that was well-known due to an incident that occurred on November 7, 1916, when it announced the election of Charles Evans Hughes as president, only to be contradicted by the next morning's newspapers.[124]

To show their usefulness, American amateurs attempted to complete an exceptional collective project. In February 1917, the American Radio Relay League proposed to amateurs through the magazine *QST* to volunteer for the defence service; the idea was to arrange a network of working radio stations that could spread the news and other useful information throughout the country and, therefore, demonstrate the value of radio amateurs. It was designed as a continuation of a successfully executed project, 'the Washington's Birthday test,' on relaying a message to the entire country (Figure 19).

The establishment of a regular network for relaying messages would, it was reckoned, determine whether the League would be closed down in times of war or if it could be of real use to the nation.[125] The realization was simple; each radio operator who participated in the project was asked to forward messages at a specific time for 15 minutes each day according to a common timetable. In practice, he was receiving them and then transmitting them further, with all operators advised to have an assistant to substitute for them in case of emergency so as not to break the communication channel. Several experimental tests showed that the system worked with only a few mistakes, with the amateurs encouraged to practice more and looked forward to new tests to improve their performance. The amateurs wrote that without them, the 'Navy would lose these one thousand listening ears.'[126]

When the US declared war on Germany in April 1917, the amateur journal *QST* published several articles expressing the fear that all radio might be closed down; there were 5,425 licensed radio station operators at that time. In the next

123 "The Quenching of Sayville," *The Wireless World* (November 1915): 516.

124 Jesse Walker, *Rebels on the Air: An Alternative History of Radio in America* (New York and London: New York University Press, 2001), 24.

125 Edgar Felix, "Department of Defence," *QST* February (1917): 20–21.

126 "War?", *QST* March (1917): 29.

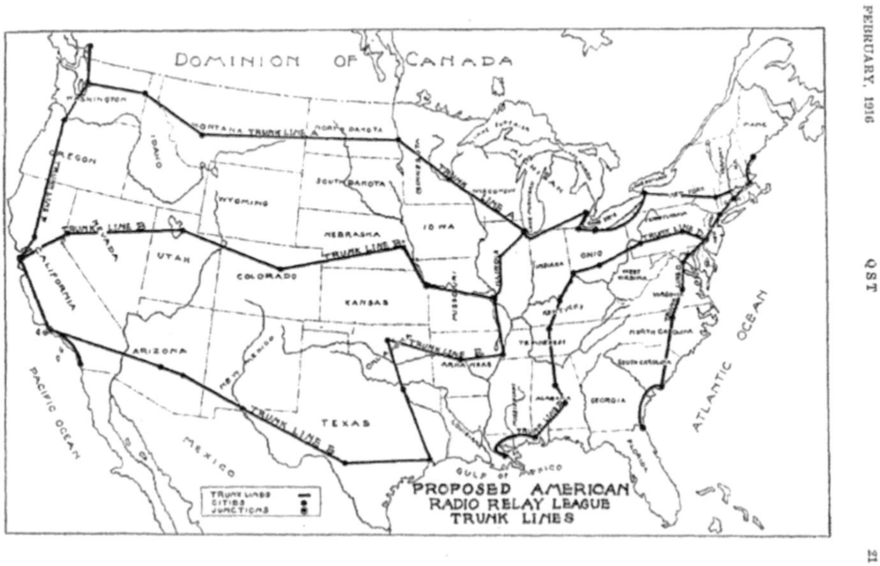

Figure 19: Relay network that American radio amateurs tried to implement in 1916–1917.[127]

issue, which was published in May, *QST*'s leading article began by quoting an order from the Department of Commerce 'To All Experimenters.' This order directed them to close all stations for radio communications (both transmitting and receiving), to lower all aerial wires down to the ground, and disconnect all devices. All efforts to convince the department of the usefulness of amateur radio in times of war had been to no avail, and the radio amateurs' network came down. *QST* described it as follows: 'All plans for improvements are cancelled. All the plans of our manufacturers are in mid-air.'[128] However, local radio clubs became more determined than ever to stay together, so as to be ready to operate again. At their regular meetings, there were discussions as to whether it was a violation of the law to leave an aerial standing if it was being used to fly the American flag. However, an atmosphere of extreme patriotism and xenophobia arose, and very soon neighbors were spying on each other, even expressing disquiet when, for example, someone installed an electric doorbell. In Buffalo, New York, the sheriff specifically set up a wireless station at the Federal building to catch amateurs'

127 "Proposed American Radio Relay League Trunk Lines," *QST* February (1916): 21.
128 "War!," *QST* May (1917): 1–3.

signals and identify illegal radio communication;[129] the next day, the newspapers reported that more than three hundred radio stations had been dismantled. This method of finding amateurs was later deployed in other cities. In 1917, American radio amateurs thus experienced many of the same governmental restrictions that had already been in place in Europe for three years, thereby affecting the course of radio development.

Educating new users: Soldiers, amateurs, children, and women

The war was also remarkable for the progress of wireless telegraphy, given the advent of new users of the technology. Apart from technological progress and innovation, mobilization necessitated the use of all possible human resources. While civil radiotelegraphy was mostly suspended and was being operated only in secret, the miliary had to train some personnel for radiotelegraph operations. First, the armies organized training sessions to teach officers and soldiers the basic principles of radiotelegraphy, then, radio amateurs were involved in different divisions as technical experts who would teach their colleagues how to use the new telecommunication tool. Finally, women and children also proved to be useful in performing basic radiotelegraphic tasks.

Military necessity also obliged governments to set about eradicating illiteracy as regards radio. Even though various states had done much to develop radio communication and to acquire radio equipment, many troops were not familiar with the technology. Only professional wireless telegraph workers could claim to have undergone consistent professional training, and these were very rare. Even deployed radio amateurs had received a very fragmented education, mainly of secondary literature and journal articles or at best general physics courses.[130] In the first months of the war, it became clear that commanders and even officers did not know the essential features of the radiotelegraph, such as its advantages and pitfalls, and in most cases could not correctly use it to control troops. One of the most striking examples of the underestimation and inappropriate use of radio communications for command and control was the defeat of the Second Russian Army commanded by General Samsonov in East Prussia in August 1914. While pursuing the enemy, the military unit suffered from poor organization of communication with the other corps. Not knowing or perhaps even disregarding the par-

129 "A Brief Account of What Happened at Buffalo . . .," *QST* 7 (1917): 26.

130 Jessop, *The Bright Sparks of Wireless*.

ticulars of radio communication, radio operators transmitted a number of operational documents without encryption and thereby revealed the intentions of the Russian command to the enemy, which led to heavy losses. When the circumstances of the defeat were revealed, the Supreme Commander-in-Chief of the Armed Forces issued a directive forbidding the transmission of unencrypted radiograms, however, this directive caused an unexpected reaction. Encryption was considered an unwelcome complication, and radiotelegraphy in general a risky communication instrument, which led to the overall (albeit temporary) abandonment of radiotelegraphy by Russian soldiers in the first years of the war.[131]

To eradicate illiteracy in radio communication, many countries organized various rush courses and training teams, so as to acquaint personnel with the new technologies. For instance, the British Royal Fighting Corps set up specific training that took the usage of radiotelegraphy on planes to an entirely new level. Before the war, a pilot who could fly a plane one kilometre and then successfully land was already considered to be useful. However, during the war, pilots became more skilled by learning general discipline, military law, the use of machine guns, the theory of flight, and map reading. Among these skills were also nuances of wireless signalling and receiving, so radiotelegraphy became an essential part of a pilot's education.[132] In Russia, the Electrotechnical School of Officers called up officers from the reserve who had had some education in the fields of electricity, and engineers, such as those who had graduated from a military engineering school, with these officers trained in the practices of radiotelegraphy and later sent to the front. The radio divisions of the Northwest, West, and South-Western Fronts also organized crash courses for the junior radio specialists, such as radio operators and electromechanics.

Help in introducing the new technology came from enlisted radio amateurs, who were one of the most important social groups to use the radiotelegraph during the war. On the one hand, they shared their knowledge and experiences to help others better understand the characteristics of invisible communication waves, while on the other hand, amateurs also learned new skills in engineering and in the day-to-day practice of technologically equipped units; they applied their knowledge to real-life tasks, underwent specially organized training, and experimented with technology.[133] Furthermore, some gained access to state-of-the-

131 Alexander Ivanovich Andogsky, *Meeting Engagement. Strategic and Tactical Investigation of Ways and Methods of Conducting Meetings of a New Battle in the Modern Era on the Basis of Military-Historical Examples [Vstrechnyj Boj. Strategichesko-Takticheskoe Izsledovanie Sposobov i Priemov* (Petrograd: General Staff Academy, 1918).

132 "Training Aviators," *The Wireless World* (July 1917): 234–35.

133 Bruton and Gooday, "Listening in Combat – Surveillance Technologies beyond the Visual in the First World War."

art devices, as states invested heavily in technical equipment. Overall, amateurs also contributed to the progress made in the course of the war; World War I had become an opportunity for some amateurs to progress using military equipment and national resources so as to address the urgent need for information. However, the exact number of amateurs who served during the war is not known because information about private stations was considered too sensitive to disclose. Even in the UK, where the radio community was the most institutionally developed, there is no evidence extant because the publication of the national Directory of Amateur Wireless Stations was postponed in order to keep all the names, addresses, and other meaningful information related to amateur stations secret.[134] It is known, however, that on March 31, 1914, 1,963 licenses were issued.[135] In other countries, such as Germany and France, societies and clubs preferred to operate autonomously and regionally and did not collect figures covering the whole nation.

Amateurs played a significant role in introducing broadcasts to the military. For instance, in Germany, Hans Bredow held several governmental positions within the postal affairs ministry as an electrical engineer and is credited with having arranged one of the first experimental music broadcasts to troops at the front, which took place in 1917.[136] Furthermore, amateurs drafted into military wireless operations spread the term 'broadcast' to describe sending messages to many naval vessels at once.[137]

Apart from military schools and the direct involvement of technical enthusiasts, governments also encouraged the sharing of new information that could lead the country to the victory. In fact, amateurs played an important role in the field of research; they thus reacted and responded to the latest technical achievements. Thus, on October 1, 1915, Moscow Military-Industrial Committee organized the Department of Inventions which aimed to collect information regarding the most recent inventions, to be informed of the needs of the army and to organize the mass production of inventions with the involvement of scientific work institutes and military plants. During the next year and a half, 1,243 suggestions for possible inventions were made, and the Department sanctioned further development of some of them. Radio amateurs developed a mechanism for pulling bombs out of airplanes, an automatic gun and an explosive bullet to fire at airplanes, as

134 "Notes of the Month," *Wireless World* (1914), 372–73.

135 "Notes of the Month," *The Wireless World* (January 1915): 640–41.

136 Gilfillan, *Pieces of Sound. German Experimental Radio*, 33.

137 P. Satia, "War, Wireless, and Empire: Marconi and the British Warfare State, 1896–1903," *Technology and Culture* 51, no. 4 (2010): 852.

well as a relay for the phonetic phone.[138] The requests and reviews of radio amateurs thus also represented a valuable source for those studying how the radio field was understood by political actors.

Amateurs were keen to be involved in the war effort. Portuguese amateurs helped the British Navy, including Alberto Carlos de Ilveira, who assisted the British Navy from Cape Verde.[139] Moreover, they maintained transnational relations, and inventions were also discussed transnationally. For instance, it was reported that Mr de Lange in Britain claimed to have succeeded in making a thermophone, with his success inspired by the experiment of an unnamed Russian, who obtained a good thermophone by placing a Wollaston wire in an insulating medium and then treating the wire with acid.[140]

The application of radiotelegraphy to military aviation was an important step in the spread of radio. During the war, various countries attempted to install radio sets in airplanes, and a special military department created in France hired radio amateurs and physicists able to offer unique expertise in this young field. For example, Edmond Rothé, a professor of physics at the University of Nancy, left his fingerprints on the area of aviation as a founder of a university institute of aerodynamics and as the technical secretary of the Society of Friends of Aviation (SAA). His main achievements concerned the application of radio to aviation. During World War I, he joined the radiographic service of the military hospital in Nancy and adapted the wireless to planes of the 1st Arms Corps. Although never viewed by historians as a pioneer in this new media, Rothé in fact did much to change the perception and uses of radio in aviation, addressing and overcoming the most pressing problems arising out of the development of these technologies.[141]

In France, the Société française radioélectrique (SFR) had been very active in the service of the army during the war.[142] Even children were taught to help with telephone, telegraph, and radio work. In Paris, boys from 10–16 years old from the scout organization Les Eclaireurs de France were trained by the Department of War and the Department of the Navy to be ready at three hours' notice to act

138 "Chronicle [Hronica]," *Herald of Military Radio Telegraphy and Electrical Engineering [Vestnik Voennoj Radiotelegrafii i Jelektrotehniki]* 3 (1917): 132–33.

139 Silva, "From Point-to-Point to Mass Communication: The Radio in Portugal from 1898 to 1939."

140 "Results of Russian Experiment," *The Wireless World* (February 1915): 686.

141 Laurent Rollet and Philippe Nabonnand, "Why Aerodynamics Failed to Take off in Nancy: An Unexpected Casualty of World War I," in *A War of Guns and Mathematics*, ed. David Aubin and Catherine Goldstein (Providence, R.I: American Mathematical Society, 2014), 351–69; Arnaud Saint-Martin, "L'office et le télescope: une sociologie historique de l'astronomie française, 1900–1940" (Université de Paris-Sorbonne, 2008).

142 Duval, *Histoire de La Radio En France*.

as messengers, to assist post offices flooded with soldiers' telegrams and individual delivery letters.[143] Meanwhile, the American QST was recruiting amateurs for the armed services. The July issue had the title, 'Wanted by Uncle Sam: 2000 amateur wireless operators' and included letters from amateurs describing their experiences at military training stations and an application form. In the summer of 1917, Harvard University offered a radio school, and a very similar school was established on the Pacific Coast at Mare Island. Nearly 5,000 radio students were attending these classes by December 1917, a number that of course does not include the majority of amateurs then serving in the armed forces and undergoing military training.

The war also had a major influence on negotiating the role of women in the development of wireless in Europe. In the absence of men, women began filling the men's positions and proved to be quite successful and efficient. In the UK, there was even a particular organization formed: the Women Signallers' Territorial Corps, which was described as 'undoubtedly the most useful and effective of all the semi-military organizations of women,'[144] with the main goal to release men from signalling work so that they could join the army. Wireless telegraphy was one of these particular fields, and the women of this organization were eligible for a course organized by the East London Wireless College. In the midst of war, the Women Signallers' Territorial Corps was so keen to learn wireless telegraphy that it asked the Marconi Company for a supply of the apparatus. The politicians also supported their work, noting: "The government has made it clear that it is incumbent upon women to fill every possible post that may be vacated by men fit for service in the forces."[145]

The Women Signallers' Territorial Corps had also suggested employing competent operators at postal telegraph and camp telegraph offices,[146] with the British Government appointing women in charge of the wireless stations at Rathlyn Island and the Island of Mull.[147] The newspapers and journals noted the following: "This real attempt on the part of women to acquaint themselves with some of the world previously undertaken by men deserves the highest commendation."[148]

After having been accepted as wireless operators, women were also recruited for constructing and testing the apparatus for the aerial forces.[149] Women thus

---

143 Felix, "Department of Defence."
144 "Wireless Telegraphy for Women," *The Wireless World* (February 1916): 725–26.
145 Ibid.
146 BT Archives, "POST 30/2589B. Employment of Women as Wireless Telegraph Operators. 1913–1920" (1915).
147 "Our Article in the February Number . . .," *The Wireless World* (March 1916): 793.
148 Ibid.
149 "More Women for Wireless," *The Wireless World* (February 1918): 768.

Figure 20: Women at the instruments, 1916.[150]

undertook an important role in many wireless departments (see Figure 20), with one notable exception: on board ships.[151] In the UK, the Women's Royal Naval Service (WRNS) started hiring women from 1917 but only as domestic wireless operators.[152] Maritime wireless telegraphy was still seen as unsuitable, women still not being trusted in times of crisis. The note in *The Wireless World* of 1918 had asserted that 'the feminine temperament is an uncertain factor in times of emergency' because there is always 'the likelihood of her natural weakness revealing itself at a critical moment.'[153]

Conclusions

The outbreak of World War I immediately imposed particular challenges for the radiotelegraph in the international arena. To the warring parties, the radiotele-

150 "Wireless for Women."
151 "Wireless for Women."
152 Elizabeth Mary Bruton, ""Uncertain at Present for Women, but May Increase": Opportunities for Women in Wireless Telegraphy during the First World War," *Information & Culture* 55, no. 1 (2020): 51–74, https://doi.org/10.7560/IC55104.
153 "Marconi "Amazons"".

graph was no longer the subject of international regulations. Decisions that had been made in the 1910s were promptly set aside, thereby reopening the period of interpretative flexibility for the radio. The military began using radiotelegraphy exclusively as a point-to-point channel, prioritizing the secrecy of the information transmitted. In some particular situations, radiotelegraphy became the only possible communication channel between allies; it was still a transnational tool, but it was no longer intended to be shared by many nations and instead the paradigm was developed of an exclusive telecommunication. The hostile situation in Europe also led to all international conferences and collaborations being deferred, rendering it impossible for the ITU to function as an international arena. Furthermore, collaborations in the war zone also stimulated the transfer of technology and knowledge. Non-European powers likewise experienced important technological advances; some were actively exploited during the war, and others acquired new resources and technologies.

Radiotelegraphy was most often used to supplement electric telegraphic communications throughout the war. However, there were particular cases where it surpassed all other means of communication, due to its having at least three unrivalled advantages. First of all, it occupied the most important position in those territories where cable communication had been intentionally destroyed either by the enemy or to prevent use by the enemy. As many cable communications were liquidated during the war, ever larger tracts of territory were covered by the invisible and immaterial waves of radiotelegraphy. Second, radiotelegraphy was essential for mobile communication. Along with ships at sea, military divisions began using the radio while on the move. Radio stations were routinely installed on buggies and then gradually moved to the airplanes. Third, radiotelegraphy served as an important tool for transnational communication. Its capacity to easily transcend national borders helped combatants to reach distant places and to enter into contact with other armies, with these military divisions thus offered a unique opportunity to connect with each other though far apart. These three features of radiotelegraphy (immateriality, mobility, and extreme distance) were the most important factors for the use of radiotelegraphy in combat.

During the war, the wireless companies also served national interests. Their resources were fully mobilized and served to supply the armies with the equipment required. Military needs for particular apparatuses determined the production of the entire industry for several decades after the war, while the necessity of using radio in combat highlighted the requirement for portable communication devices. The need to also employ them on aircraft also determined the requirement of being portable and lightweight, while the need to communicate quickly also led to advances in voice transmission as well as in code. This demand created a supply, which constituted a path dependency for radio technology. Moreover, women who had previously been discouraged from entering the workplace proved to be an indispensable human resource in organising radio transmissions.

The transnational networks were disrupted. Few stations were regularly sending out the time signals and weather reports, as the armed forces were exploiting most of them; this disruption in the networks also rendered these services unreliable or even no longer meaningful. Voice and music broadcasting stations were shut down altogether following the outbreak of the war, however, newspaper agencies were more fully employed than before. Even though they also were censored and forced to work for the military, they remained one of the very few information sources and sometimes were also used for counter-propaganda.

Most European countries banned individual stations after the outbreak of the war, in order to avoid the risk of espionage. However, some amateurs were still operating illegally, and offered their services to the nation. Others proved their usefulness in the war in radiotelegraph divisions and other technical departments, yet others provided officials with necessary information regarding incoming radio messages. Many, however, were not put to any use in the war, the overriding concern being to keep military information secret. For this the community of radio amateurs undoubtedly suffered.

World War I may thus be said to have presented radiotelegraphy in a new light: as a point-to-point tool whose openness could be used for great effect in particular situations. The transnational dimension of radiotelegraphy also acquired a different meaning; it was still intended for communication between nations, but only among privileged actors.

Chapter 4
Establishing a new order, 1918–1927

The war had drastically changed the social and political landscape. It is tempting to interpret the rise of radio broadcasting in the 1920s as an immediate and irreversible consequence of World War I, however, many decision-makers, such as engineers and managers in governmental bodies and telecommunication companies, had sought to retain the management policies for wireless that had been agreed upon in the early 1910s. The change as regards radio did not happen immediately but was instead a gradual, indeed, inevitable transformation throughout the 1920s. It began with the first attempts to adjust those international regulations that had proved most controversial and led to the complete reconfiguration of the role and place of radio in society.

This chapter discusses how several attempts to organize flawless radio communication in Europe were made immediately after the war at numerous international meetings and events. It takes into account the ITU actions in this period and the creation of new international organizations that endeavored to regulate particular aspects of radio. It also analyzes the 1927 ITU Conference in Washington, which drew a line under the different interpretations of and perspectives on radio and wireless. The chapter then goes on to focus on the new companies that opened in the sector of public radio broadcasting and the challenges they posed within the international arena. In addition, this chapter examines transnational networks through the maps of the radiotelegraph and time and weather services. Finally, it concludes with a discussion of the users and amateurism itself, including regulations governing ham radio.

Attempts to restore the global wireless networks

International arenas after the war

In June 1919, the American journal *The Wireless Age* had noted that the war had contributed greatly to the development of wireless worldwide. A new term highlighted the significant spread of infrastructural networks around the globe: the World Wide Wireless. The article also referred to the 'copyrighted dispatch' from London on Marconi that had expressed the hope that the developments in wartime would finally contribute to the realization of the vision of embracing the whole world and connecting the globe. From waterways and aerial routes to the most remote places in the

Open Access. © 2024 the author(s), published by De Gruyter. BY-NC-ND This work is licensed under the Creative Commons Attribution-NonCommercial-NoDerivatives 4.0 International License.
https://doi.org/10.1515/9783111202327-005

world – everything was believed to have finally become a part of a global communication network.[1]

However, the rapid and incessant development of radiotelegraphy, as well as the technical progress achieved in the last years of the war, had shown that the existing international regulations were no longer adequate. For example, call signs, which were formed using combinations of three letters according to the 1912 regulations, were notoriously insufficient, as the number of radio stations had grown exponentially. Some offices found themselves obliged to resort to the use of four-letter codes, even though it was not prescribed in the existing international regulations.[2] This and other changes in radio required a complete revision of the existing international agreements.

The ITU network tried to consolidate itself immediately after World War I to solve the problems arising as a consequence of the war. In 1918, favourable dynamics had already emerged regarding international collaborations. In correspondence with the ITU, many countries began requesting information about international affairs, which led to a rise in the amount of correspondence. In the light of this growing interest in international negotiations, the editors of the *Journal Télégraphique* predicted that new international meetings on radiotelegraphy would no longer be delayed: 'During the present year, when peace is finally concluded, Administrations without a doubt will be concerned with the meeting of the various congresses and conferences, postal confederations, telegraphic conferences, radiotelegraph conferences, which the international political situation had postponed.'[3] However, it would be nine long years before an international conference was held, in Washington, in 1927.

That large international conference had been preceded by numerous smaller gatherings, which together shaped the new understanding of radiotelegraphy. The European countries initiated these gatherings, especially Great Britain, France and Italy, with extra-European players, such as the US and Japan, later included. With new actors emerging from different parts of the globe and new visions arising for the technology, the agenda at every meeting drifted further and further away from the ideals of the 1910s. The attempts to restore an accepted understanding of global communication led therefore to the establishment of an entirely new order.

The first attempts to discuss the future of radio arose as early as 1919. The war had alerted many to the need for an international code of signals, like those accepted by the navigation companies. Therefore, in November 1919, representatives from

1 "World Wide Wireless," *The Wireless Age* 6, no. 9 (1919): 7–10.

2 "L'Union Radiotélégraphique," 154.

3 "Revue Télégraphique de 1918," *Journal Télégraphique* 1 (1919): 1.

Great Britain, France and Italy gathered in London to discuss the feasibility of establishing an agreed set of international codes.[4] This meeting remained an unofficial gathering of various state representatives but showed how urgent it was to establish a more detailed and nuanced international scheme of radio communication.

On July 7–13 1920, Paris hosted the meeting of the European administrations to discuss international communication. This European conference on international communications touched upon the issues concerning radiotelegraphy that would define the international agenda of later international conferences managed by the ITU. At this meeting, the administrations in attendance considered radiotelegraphy to be a transnational technology, almost as it had been in the 1910s. Radiotelegraphy was seen as the most used telecommunication channel, especially for international communication, and was acknowledged as 'the fastest and therefore the most frequently used route' for communication. In particular, international communication was 'increasingly preferred by radiotelegraph.' Moreover, the use of a land route, such as a wired electric telegraph, was even called 'exceptional.'[5] One of the issues discussed was whether the sender would have the opportunity to indicate their favored means of telecommunication when sending a message. The initial proposal was to have a particular mark, namely, 'radio' if one preferred to use the radiotelegraph, but the delegates decided that instead of 'radio,' it would be more reasonable to adopt the mark 'wire,' as wired communications were rarer.

In the autumn of 1920, at the invitation of the US government, representatives of the US, UK, Italy, Japan, and France held a Preliminary International Conference on Electrical Communications in Washington, D.C. The initial aim was to determine the fate of the German cable lines that had been ceded to the Allies in accordance with the Treaty of Versailles in 1919.[6] Moreover, the agenda covered a number of more general problems of international communications, including those regarding radiotelegraphy. The ITU Bureau had contributed to the preparations for the meeting by sharing materials on radiotelegraphy that had been drafted or collected for the 1917 Washington conference that never took place.[7] The name chosen for the meeting (*Electrical communications*) indicated that all communications should be treated equally regarding their contribution to the international communications network. Its agenda suggested discussing the im-

4 "Code Radiotélégraphique International," *Journal Télégraphique* 11 (1919): 183.

5 As cited in ITU Archives, *Documents de La Conférence Télégraphique Internationale de Paris, 1925, Tome II (Paris, 1925)* (Bern: Bureau International de l'Union télégraphique, 1926), 240–41.

6 National Archives of the United States, *43.2.11 Records of the Preliminary International Conference on Electrical Communications* (1920).

7 "Inauguration Du Monument Commémoratif de La Fondation de l' Union Télégraphique," *Journal Télégraphique*, no. 12 (1922): 243.

provements that should be made 'to contribute to the organization and operation of the global electrical communications network.'[8] Therefore, at this conference, radiotelegraphy was treated not as a unique means of communication but instead as an integral part of a global electrical communications network.

Furthermore, this conference prompted the idea of merging the two international arenas of the ITU on telegraph and radio into one. In doing so, it was proposed to establish an entirely new organization – a Universal Electrical Communications Union. The name chosen had been intended to echo that of the Universal Postal Union, thus highlighting certain similarities between postal and other communication services. The delegates proposed that discussions on telegraphy, telephony, and radiotelegraphy should presuppose a shared arena, and that the management of the telecommunications sector should everywhere be subject to the same general regulations. As a result of this meeting, a draft convention was drawn up that contained in one act all the relevant provisions: those for both the telegraph and the radiotelegraph.[9] The key figure promoting this vision of a single organization for the world's cable and radio networks was Walter S. Rogers, whose career and role had been somewhat neglected until the recent research done by Richard R. John. While in the short run, his project and the conference should be judged to have failed, in the longer term they drew attention to certain limitations of liberal internationalism, and later, at the ITU conference in 1927, empowered national governments to determine how their portion of spectrum would be used.[10]

The subsequent meetings took place in Paris and Riga in the summer and autumn of 1922, respectively. A group of experts deepened this international collaboration and discussed future, putative regulations in greater detail, with the results of their discussions passed on to the ITU Bureau. In all their proposals, radiotelegraphy was considered in relation to the international telecommunication network. One of the questions addressed the attempts made to include Russia in the global system of communications, with radiotelegraphy operating fees being at issue.[11] Though occupying large territories and despite having been a prominent actor at international meetings in the past, Russia was left outside of the international arenas at this juncture, due to its changed political situation.

---

8 "Revue Télégraphique de 1922," *Journal Télégraphique* 1 (1923): 3.

9 "Conférence Internationale Sur Les Communications Électriques," *Journal Télégraphique* 12 (1922): 259.

10 Richard R John, "When Techno-Diplomacy Failed: Walter S. Rogers, the Universal Electrical Communications Union, and the Limitations of the International Telegraph Union as a Global Actor in the 1920s," in *History of the International Telecommunication Union* (2020), 55–76, https://doi.org/10.1515/9783110669701-004.

11 "L'Union Radiotélégraphique."

In 1922, the ITU Bureau sent out the draft for a new convention to all administrations, who for their part were invited to draw up proposals for any modifications that they wished to introduce. Other meetings were also expected to follow, however, the meetings discussing these same suggestions continued to be held outside of the larger international ITU arena. In April 1923, Mr. Salandra, the representative from Italy to the Council of the League of Nations, reached out to the League of Nations with a memorandum in which Italy persistently called for the convening of an international radiotelegraph conference:

> In the opinion of the Italian Government, in the presence of the extensive development of the applications of science to radiotelegraphic and radiotelephone transmission, it would be fitting to lay down international rules to specify the rights and duties of the Companies and Governments.[12]

The Council of the League of Nations decided to have the Italian memorandum scrutinized by a committee of specialists consisting of the directors of the British, French, and Italian telegraph services. They met for the first time on July 16 in London at the Office of the League of Nations. At this meeting, the delegates agreed that a universal international conference should be convened in order to resolve how both the telegraph and the radiotelegraph might be managed. However, it was noted that the French government was already organizing an international conference on telegraphy, which could be used as an appropriate arena to discuss these very questions. On November 13, 1923, the same British, French, and Italian experts met once again in Geneva, agreed that the international conference on radiotelegraphy should be held as soon as possible, and hoped to set the date for the spring of 1924. Many documents were already prepared for the conference, such as the draft of the programme and convention, however, nothing definite had been decided as regards the future meeting. Furthermore, the ITU Bureau received no formal communication to convene the conference.[13]

These meetings were the preliminary discussions about the future of radiotelegraphy, and most of them were still suffused with the rhetoric of the 1910s. The suggestion was made to incorporate radiotelegraphy into a global telecommunications system and to build a coherent radio space. The experts were endeavoring to treat radiotelegraphy as a point-to-point communication network that could prove useful for transnational purposes, however, these transnational issues raised further problems and questions about regulation, which could hardly have been predicted in the early 1910s but became highly relevant after the war. Governments

12 "Revue Télégraphique de 1923," *Journal Télégraphique* 1 (1924): 3.
13 "Revue Télégraphique de 1923."

had learned the importance of control over the radio waves after experiencing the enforced shutting-down of stations. In particular, they were suspicious about the possible use of radio by spies and raised doubts about the transnational management of radiotelegraphy with regard to private individuals. For example, a wide-ranging discussion arose on the issue of how to deal with the licenses for reception and transmission issued to foreigners.[14] In the US, licenses for transmitting radio stations were issued only to US citizens, while in the UK, these licenses were granted irrespective of nationality.[15] The understanding of radiotelegraphy as a transnational communication space contradicted the need to restrict and limit radio communication in the case of a national emergency, and this contradiction was not easily solved.

The 1925 International Telegraph Conference

In September and October 1925, at the Paris Conference, the state representatives gathered to modify the existing set of rules concerning telegraph. As at the previous telegraph-focused meetings, the issue of radiotelegraph was also briefly discussed. The discussions that had taken place among the European countries in the early 1920s had a strong influence on the debate at the 1925 conference; delegates referred to the issues raised and materials under consideration at the previous meetings and also adopted some of the previous suggestions regarding this international arena. For instance, the 1925 conference approved the proposal made at a 1920 meeting about only marking messages to be transmitted by wire, if such a circumstance arose, as the radio seemed to be a more common and preferred mode of communication and did not require any particular mark. Therefore, the international arena acknowledged the role of radiotelegraphy as a default communication channel.

Another important point emerged from the discussions in Paris and also at further conferences: language. The delegates had used French as a *lingua franca* for many decades, however, it was becoming less and less relevant in a world that had been politically recast. The reasoning was simple; French was indeed widely spoken, but it 'did not overflow beyond its natural borders,' meaning that it was widely spoken mainly in France and its colonies. Other languages, however, were dominant even outside of their country of origin, like English. The spread of English

14 ITU Archives, "Circular No. 186," 1925.

15 Department of Commerce Washington, "D. 40. N. 25," in *Correspondance: Radiotélégraphie* (ITU Library & Archives, 1925); General Post Office London, "D. 40. N. 22," *Correspondance: Radiotélégraphie* (ITU Library & Archives, 1925).

was, it was said, 'natural, spontaneous and irresistible.'[16] The motion to adopt English as the second language of the conferences came from the Japanese Delegation, and was seconded by the Chinese, while the delegations of Germany and Ireland opposed the idea of changing the language. An Irish delegate remarked: 'French is clear enough, there is no reason for adopting a second language, which would inevitably lead to adopting a third, fourth, etc.'[17] In the end, the conference rejected the Japanese proposal, but not for long; English became a second language at the 1923 Madrid Conference.

How should we interpret this proposal to change the language? First, it demonstrates how the political and social order had changed and that the English-speaking countries had acquired particular power after World War I. Second, it indicates that there had been a crucial shift in political discourse where the presence of the extra-European countries, particularly Japan, was concerned. They were no longer dismissed as entities within an unimportant extra-European space and characterized by a rather mediocre technological development but instead had become key players with a voice. Overall, this reflects the comprehensive shift in the nature of the social groups that were shaping the development of radio.

The conference had also considered the merits of merging the two telegraphic and radiotelegraphic conventions, and there appeared to only be advantages in bringing together basic provisions about telegraphy, telephony and radiotelegraphy within a single context. Furthermore, the conference had officially set up within the ITU the International Long-distance Telephone Consultative Committee (CCIF) and created the International Telegraph Consultative Committee (CCIT). The creation of a similar committee for radio, and a number of other initiatives, formed a key part of the agenda for the 1927 Conference on Radiotelegraphy in Washington.

The proliferation of radio broadcasting stations

The Armistice of November 11, 1918 ushered in a new phase in the development of radio telegraphy in Europe, with national consolidation and the organization of a new social order the primary concerns of European leaders. Many countries were therefore concerned with establishing efficient and regular transmission of nation-building messages. Thus, at the end of the war, national governments

16 Th. Le Danic, *La Conférence Télégraphique Internationale de Paris (1925)* (Paris, 1926), 30–31.
17 ITU Archives, *Documents de La Conférence Télégraphique Internationale de Paris, 1925, Tome II (Paris, 1925)*, 153.

were in no hurry to relax the constraints on radio communication, preferring to hold on to their monopolies for a period of time.

The war raised concerns about the use of radio. Radio broadcasting spread in Europe slightly later than in the US, primarily because of the various governments' resistance on military grounds.[18] Many governments did not lift the ban on radiotelegraph stations immediately after the war, and for some years the advent of broadcasting was delayed. Even well-established companies were debarred from experimenting; in 1920, for instance, the British government banned Marconi from conducting any broadcasting experiments, it being obliged rather to serve the armed forces.[19] Before that, however, the two mammoth wireless masts of 450 feet installed in Chelmsford were frequently used for experimental transmitting work. Most notably, on June 15, 1920, they were used for broadcasting the concert of Dame Melba, more than two years before the official start of broadcasting in the UK.[20]

The spread of new technologies, the gradual introduction of radio to the general public throughout the war, and the widespread admission of the usefulness of broadcasting had nonetheless paved the way for the appearance of broadcasting stations. For the establishment of national broadcasting stations, several specific changes in the understanding of the technology were required.

First, the idea of using radio to broadcast information instead of as a point-to-point, two-way channel of telecommunication had to be accepted. Broadcasting practice, as mentioned above, originated in the 1910s and evolved from meteorology reports and time signals, which had specifically targeted many recipients at once. Indeed, meteorology and time signals constituted an important part of radio's history; 1921 is known as the year of the birth of radio broadcasting in France, as the Eiffel Tower started issuing daily transmissions in December, which included live concerts of musicians performing in a studio, and many newspapers published photographs of artists and entrepreneurs at the Eiffel Tower.[21] However, what often eluded the journalists was the fact that time signals and meteorological information also featured. In other words, these transmissions were in their form all but identical to those already taking place in the 1910s, however, the role of time signals and meteorology in the first broadcasts was underestimated from the very beginning. The transmission of the concerts was not, in fact, an innovation pure and simple

18 Andrew Crisell, *An Introductory History of British Broadcasting* (London and New York: Routledge, 1997), 17–18.

19 Williams, *Television: Technology and Cultural Form*, 27.

20 "Some Essex Industries. The Romance of Marconi's wireless wonders at Chelmsford." *Essex Weekly News*, August 23, 1935. Marconi Archive. Ms Marconi 200.

21 E.g., Yvonne Printemps, "Concert à La Tour Eiffel Par TSF," *Agence Rol* (1922).

but just a continuation of an already established practice dating from the 1910s and following the same pattern.

Second, experiments at transmitting voice and speech finally bore fruit during the war when certain devices went into mass production. Wartime research greatly enhanced the technology of amplitude modulation (AM). The spread of the vacuum tube precipitated a rise in AM radio broadcasting in the 1920s, as it was relatively cheap and also more spectrum-efficient. Furthermore, crystal transmitters led to a wholesale transformation of the industry. Even at very high spark rates, spark transmitters were too noisy to transmit speech without intolerable distortion, and this shift to AM helped boost radio voice transmission. It also contributed to tuning the radio because the signals of spark transmitters bled across the spectrum. Even though crystal had been invented at the beginning of the 1900s, it was kept hidden as a secret military innovation and only underwent applications of its full capacity during the war;[22] it would not reach the consumer market until after the war,[23] when amateur radio devices were thereby rendered more robust and easier to construct. Previously, hobbyists had tried to build similar do-it -yourself devices, but very few amateurs in Europe met with any success. For example, Horace Hurm manufactured and sold a small crystal detector, the "Ondophone," in Paris around 1914, but it only worked near the Eiffel Tower.[24] The war also mobilized commercial companies and accelerated the progress in vacuum tubes. After the production of hundreds of devices for military communication, a strong industrial base was laid to mass produce radio sets for domestic use in the 1920s.[25] This success was due to greater experience and familiarity, and to the spread of the crystal detector for wireless receivers.

Third, these new services had a major impact; the news agencies, having been exclusive holders of the news and other information being transmitted by radio, began to realize that their subscribers were increasingly able to pick up the news from the air. To optimize their profit, these news agencies started setting up their own broadcasting services. The 1921 international meeting took place in Riga, where experts discussed the future of radiotelegraphy and argued in favor of a service that would allow messages to be transmitted by radiotelegraphy to several recipients in different cities. It was decided that the general rules

22 Satia, "War, Wireless, and Empire: Marconi and the British Warfare State, 1896–1903."

23 Jennifer Spohrer, "Ruling the Airwaves: Radio Luxembourg and the Origins of European National Broadcasting, 1929–1950" (New York City, Columbia University, 2008), 40.

24 Grant Wythoff, "Pocket Wireless and the Shape of Media to Come, 1899–1922," *Grey Room* 14, no. 51 (2013): 43, https://doi.org/10.1162/GREY_a_00106.

25 Balbi, "Wireless's "Critical Flaw": The Marconi Company, Corporation Mentalities, and the Broadcasting Option," 16.

on radiotelegraphy should also be applicable to this service, including the fees levied for using the communication channels and other general provisions. In these discussions, radio broadcasting had remained a topic of lesser importance, but this did not last long. Once again, radio encountered the problem of names; experts did not yet know how to refer to broadcasting. At the meeting in Riga, they called this service the 'circular telegrams of the press.'[26]

The shift to radio had not happened rapidly. The transition from wireless to radio was a gradual one in the popular consciousness too, as may be gathered from the change of content in the songs devoted to this means of communication. A representative example is the song, 'Kiss me by wireless,' which was published in 1922 in the US.[27] The song was distributed not only as piano sheet music, as was common for songs of the 1910s and earlier, but also as an actual recording broadcast on the radio.[28] It was recorded by bandleader and pianist Vincent Lopez, whose band is believed to have been the very first to broadcast live.[29] The title of this song does itself evidently refer to the history of wireless; indeed, it retains the same semantics and lexicon as the previously mentioned song from 1911, 'Send me a kiss by wireless.' The lyrics likewise refer to another song from 1913, sharing as it does the same opening line, 'There's a wireless station down in my heart.' Two of these references are indeed so similar to the lines of the songs from the 1910s that today they might even be considered outright plagiarism. The lyrics also mention the sea, which was another key topic for wireless telegraphy, and the overcoming of a long distance in the relationship between a couple.

The cover illustration suggests that the characters are able to stay in touch despite being in startlingly different locations: military personnel posing against a tropical background and a lady in front of a snowy landscape (Figure 21). However similar it may have been to other musical content, this song also has unique features that reflect the gradual transition to radio broadcasting. First, the message is not associated with a physical and written message, as it had been in the 1911 song. Instead, it is represented on a cover as two lines connecting the lovers. At the very end of the song, the word 'wireless' is replaced with 'radio,' and the last line states, 'For the radio station's my heart.' The recording was indeed played widely

---

26 "L'Union Radiotélégraphique," 125.

27 Justin Ring, *Kiss Me by Wireless* (Ring-Hager, 1922).

28 DAHR, "OKeh Matrix S-70586. Kiss Me by Wireless / Hotel Pennsylvania Orchestra; Vincent Lopez," Discography of American Historical Recordings (1922), https://adp.library.ucsb.edu/index.php/matrix/detail/2000202719/S-70586-Kiss_me_by_wireless.

29 Jim Cox, "Vincent Lopenz. The Piano Kid," in *Musicmakers of Network Radio: 24 Entertainers, 1926–1962* (Jefferson: McFarland, 2012), 210–24.

on radio broadcasting stations in America, and the disc was advertised as an 'original radio foxtrot.'

Figure 21: Sleeve cover for the song, 'Kiss me by wireless,' 1922.[30]

The world witnessed a veritable radio boom. By 1925, nearly every European country boasted a national broadcasting service and corresponding regulations. Radio emerged as a newly developed medium that would revolutionize communications. The radio was considered a national medium. Radio, once born to be a global technology, had become a national or even a local concern.

30 Justin Ring, *Kiss Me by Wireless* (Ring-Hager, 1922).

The formation of the International Broadcasting Union

Discussions about the future of radio continued in various political arenas for at least the first half of the 1920s, although no international conference specifically addressing issues raised by radiotelegraph would be convened in the early years of that decade. As a result of the new understanding of radio and the ITU's tardiness in convening its own international conference, a new international organization appeared. With the rapid spread of radio broadcasting stations across European countries, it was important to tackle the issue of the shared radio spectrum efficiently and without delay.

On April 15, 1925, the ITU received a letter from the newly opened *l'Union International de Radiophonie*, known in English as the International Broadcasting Union (IBU), which gave notice of the creation of the IBU. The new organization would assume responsibility, it declared, for the regulation of public transmissions, and it lost no time in announcing two upcoming conferences in London and Geneva.[31] In the absence of any other international conferences in the field, and due to a dearth of up-to-date information from the ITU, the IBU had the opportunity to take its place in the international arena.

In 1925, the Radio Station Marconi sent a letter to the ITU requesting an appointment with Arthur R. Burrows, in order to discuss the IBU.[32] The meeting took place on June 26, 1925, its agenda being dominated by a dialogue concerning the reasons for organizing the IBU Conference. A memorandum drafted by Capt. Peter Eckersley prepared the ground for the discussion,[33] with the primary concern being interference in national broadcasting services, which could be avoided if some stations changed their wavelengths. The memorandum described radio as a communication tool that should be treated as being of international importance, not with a view to creating a united global communication space but rather to properly organize separate national services. Once again, it is puzzling in retrospect to observe the influence of American radio upon the international decisions taken regarding the future of wireless. Despite European wireless being able to call upon a long history of its own, and notwithstanding its concern to preserve European diversity on the continent, the regulations stemmed from the example of the US. The American experiments, taken as guidance for the changes to be made, had established that some 1,500 miles was the minimum distance required between two radio stations to

31 Office International de Radiophonie, "D. 40. N. 5," in *Correspondance: Radiotélégraphie* (ITU Library & Archives, 1925).

32 Radio Station Marconi, "D. 40. N. 6," in *Correspondance: Radiotélégraphie* (ITU Library & Archives, 1925).

33 Eckersley, "D. 40. N.8. Situation in Europe as Regards Wave-Lengths."

avoid interference. The existence of 60 stations in Europe was acknowledged and mapped on to a quite complicated scheme, with it proposed to equip them with wave meters that would be calibrated using uniform calibration signals sent by the BBC (see Figure 22).

Figure 22: Broadcasting stations in Europe, 1926. This map was submitted as an appendix to the documents on the formation of the IBU.[34]

The ITU expressed some reservations regarding these proposed innovations, simply acknowledging their receipt of the letter and its possible future publication in the *Journal Télégraphique*. In the internal documents, however, the issue of broadcasting received much more attention. In 1925, the ITU radio correspondence register introduced for the very first time a new and separate category explicitly devoted to radio broadcasting as such. Previously, the topic of broadcasting had been subsumed within the other categories, such as those covering the inauguration of radio stations or modifications in the current infrastructure. That year was a turning point inasmuch as broadcasting as a topic came to be deemed worthy of separate treatment. What was addressed under this new heading had to do in the main with matters relating to the IBU.

34 Union International de Radiophonie, "Stations de Radiodiffusion Emettant Sur Des Longueurs d'onde de 200 a 600 m," in *Correspondance: Radiotélégraphie. 40.* (ITU Archives, 1926), 30.

The IBU proclaimed as its aim the establishment of a connection between different broadcasting companies from Europe and potentially from other continents.[35] The IBU was seen as a Society of Broadcasting, and the BBC took a leading role here, however, there was still a problem with the definition of 'radio.' Radiotelegraph, radiotelephony, radio broadcasting, and other radio communication – what exactly was the technology that this organization wished to regulate? Apart from the aspiration to facilitate communication between commercial companies, the Union of 'radiophony' was hard pressed to define what the field it covered actually was. The IBU had an original name in the French *L'Union Internationale de Radiophonie*, so it used the term 'radiophony' to describe this new practice. It is still more striking to note that the document that set out the basis for this Union, Eckersley's memorandum, did not even mention words like 'radio' or 'broadcasting,' specifically avoiding any names and referring to radio only as the 'wavelengths.'[36]

The 1927 Washington Radiotelegraph Conference

Finally, the first international conference about radio regulations to be convened after World War I took place in Washington, D.C., in 1927, ten years later than initially planned. The fact that it did not take place right after the war, even though many countries were requesting it, could be interpreted as a change of paradigm. In the war's aftermath, governments struggled to manage the radio industry. The very definition of radio was once again in question, not least because of the rapid growth of radio broadcasting. International collaborations jeopardized national security. New technologies and devices had been coming on to the market at a steady rate, resulting in a constant reappropriation and reinterpretation of the technology. Moreover, the reputation of the ITU was affected by its inability to provide requested information promptly and its failure to keep pace. It did not receive any notifications about the opening of most radio stations; their communications were therefore not covered by the Radiotelegraph Conventions.[37] The world had changed all too quickly.

The new convention had encompassed the many fears and doubts occasioned by the development of radio and reflected most of the changes in the industry. Three particular features are essential for understanding how the conference shaped the future understanding of radio.

35 Office International de Radiophonie, "D. 40. N. 5."
36 Eckersley, "D. 40. N.8. Situation in Europe as Regards Wave-Lengths."
37 "Revue Télégraphique de 1923."

First of all, this conference drew a line under the period in which radio had been interpreted in a great number of different ways, it having been in other words a stage of interpretative flexibility. One of the most important regulatory steps concerned the actual definitions of the field. The very first article of the Convention, and the most important one at that, introduced strict definitions of the phenomenon. This article was so important, and also so controversial, that a special subcommittee evaluated the terminology that was used most often throughout the plenary meetings concerned with the drafting of this article. Initially the article was called 'article zero.'[38] The first lines of this article introduced the most important term, 'radioelectric communication' or 'radio communication' as the wireless transmission of writing, signs, signals, facsimiles, and sounds of all kinds by means of Hertzian waves.[39]

Further definitions included the following terms: radio station, fixed station, mobile station, land station, mobile service, and others. There were some important differences between the wording employed at previous conferences and the conventions discernible in this new document. First, the entire notion of radiotelegraphy was no longer related to the sea. Even the definition of a mobile station did not mention ships at all, despite the fact that in the 1910s sea vessels were almost the sole vehicles to be equipped with mobile stations. The convention had also introduced the term 'aircraft station,' and by using the expression 'on or over the high seas' also indicated the possibility of placing a radio transmitter on a range of different air vessels.[40]

The convention also proposed a definition of 'international service.' Radio communication across national borders or in neutral international territory, such as a sea, was considered an international service. Furthermore, if a national radio communication service was capable of interfering with other radio transmissions beyond the frontiers of these countries, it was also considered an international service.[41] This new definition highlighted a new understanding of radio technology and radio stations. While the 1912 Conference had referred to radio stations using their location, such as 'coastal stations,' their location was no lon-

38 ITU Archives, "Documents de La Conférence Radiotélégraphique Internationale de Washington 1927. Tome II (Washington, 1927)" (Bern: Bureau International de l'Union télégraphique, 1928), 124.

39 ITU Archives, *International Radiotelegraph Convention of Washington, 1927 and General and Supplementary Regulations (Washington, 1927)* (London: His Majesty's Stationery Office, 1928).

40 Ibid., 11.

41 ITU Archives, *International Radiotelegraph Convention of Washington, 1927 and General and Supplementary Regulations (Washington, 1927).*

ger deemed important at the 1927 Conference. The range of the transmitter was considered, but its whereabouts was not.

Second, this conference demonstrated that radio development was being shaped by many new social actors, including some outside of Europe. The number of delegates and parties present at the 1927 Conference had greatly increased, so the debates actually required additional, more complex, organization and chairing. After some reconsideration of the role of the colonies, this conference also addressed the issue of voting. The great majority of states had voted for the granting of a single vote to each sovereign country, however, this simplistic solution challenged the voting regime prevailing in the Telegraph Union for more than 40 years, and consequently the power of the imperial nations, which had wielded a greater influence over the union than others. Thus, it had been vigorously combated by the imperial powers, notably France, Belgium, the Netherlands and Portugal.[42] The contracting countries failed to reach a compromise at Washington, and this issue would therefore remain unresolved until the conference in Madrid in 1932, with this tension revealing the shifts in global geopolitics. As Tworek notes, 'while voting rights might seem technical, they highlight the different approaches to representation, jurisdiction, and sovereignty within the international realm in the interwar period.'[43]

Furthermore, this ITU conference also ensured that private companies would have an important part to play in the drafting of documents.[44] Because the spectrum had begun to have economic value, it was no longer viewed as a common good to be shared by different nations but rather a common good to be sold, manipulated, and used.[45] After the 1925 Telegraph Conference, the 1927 Washington Conference had also established the International Radio Consultative Committee (CCIR) on this matter.

The third important point concerns radio broadcasting. The document of 1927 was also one of the first documents of the ITU, which separated radiotelegraphy from radio broadcasting. Article 13 of the Convention stipulated that the ITU Bureau should separately publish two different documents: a list of broadcasting stations and lists of all fixed, land and mobile stations with a call sign from the international

---

42 A Fis, "Les Conférences Télégraphiques et Radiotélégraphiques de Madrid," *Bulletin d'informations, de Documentation et de Statistique. Ministère Des Postes, Télégraphes et Téléphones* 1 (1933): 13–21.

43 Heidi J.S. Tworek, "A Union of Nations or Administrations? Voting Rights, Representation, and Sovereignty at the International Telecommunication Union in the 1930s," in *History of the International Telecommunication Union*, ed. Gabriele Balbi and Andreas Fickers (2019), 245.

44 ITU Archives, *International Radiotelegraph Convention of Washington, 1927 and General and Supplementary Regulations (Washington, 1927).*

45 Streeter, *Selling the Air: A Critique of the Policy of Commercial Broadcasting in the United States.*

series.[46] Moreover, it introduced the term 'broadcasting service,' which was defined as 'a service affecting the dissemination of radiotelephonic communications intended to be received by the public, either directly or through the medium of relay stations.'[47] Here, it is important to note the word 'radiotelephonic' as used to refer to music, sound and voice transmission. It also indicates that the separation between radio and radiophony was based on understanding the sound experience as a different practice. This discrepancy could be interpreted as a creation of radiophonic culture,[48] which had evolved from the 'beeps' of Morse-coded time signals and was being transformed into a particular listening practice.

The conference then went on to divide up the radio spectrum. Their concern here was with the use of frequencies by broadcasting stations, as they were interfering significantly with other radio communication services. All broadcasting stations working on frequencies below 300 kc/s (wavelengths above 1,000 m) were required to be removed no later than a year after the Regulations came into force.[49] The Conference had also marked a turning point in the regulation of radio for domestic use. Although very cheap and accessible, spark devices occupied a broad frequency band and therefore caused disturbance. The new convention restricted those old types of transmitters, therefore allowing more space for the new technologies.

By implementing these regulations, the ITU also restricted and determined the area of work for the IBU. Although the IBU became an important and growing organization, it did not supplant the ITU. The ITU's networking potential and its more general approach towards wireless communication remained a fundamental point of reference for any regulation in the sphere of radio. Typically, relations between the IBU and the ITU would be limited to the exchange of small portions of information; the IBU would send various reports to the ITU, while the ITU would acknowledge receipt of them and publish them in the *Journal Télégraphique*. In this sense, the ITU became a mediator of the IBU's actions with regard to a broader public. One of the outcomes of the ITU conference was the allocation of specific wavebands for broadcasting. Negotiations about dividing the sphere of influence, and focusing more on keeping it strictly controlled and regulated, were entered into. Decisions were taken without consulting the IBU and, moreover, were a disappointment for the IBU but nevertheless had to be respected. The IBU

46 ITU Archives, *International Radiotelegraph Convention of Washington, 1927 and General and Supplementary Regulations (Washington, 1927).*
47 Ibid.
48 U. Holl, "Radiophonie. Forschungen für ein kommendes Radio," *Historische Anthropologie. Kultur - Gesellschaft - Alltag*, 22(3) (2014): 426–435.
49 ITU Archives, International Radiotelegraph Convention of Washington, 1927 and General and Supplementary Regulations (Washington, 1927).

suggested an alternative but failed to implement it, due to lack of support from government officials.[50]

Overall, the regulations adopted at the 1927 Washington, D.C. Conference had a huge impact on other international agreements and regulations. For instance, regarding the question of radio and aviation, the modification of the use of wavelengths by aircraft stations implemented by the Radiotelegraph Convention at the ITU Conference in Washington, D.C., in 1927, led to the aerial Paris Convention of 1919 being amended.[51] The idea of radiotelegraphy had changed beyond all recognition: from a borderless world to national broadcasting. That was the new 'language of national responsibility'[52] that had emerged after, and because of, World War I.

Transnational networks reconfigured

Weather reports and time signals

The situation with time signals and weather reports had changed drastically since the 1910s, when the last international agreements on the matter had been signed. The international agreements and networks established in the course of that decade had borne fruit, and by 1927 the international networks of meteorological radio telegrams covered the entire globe. Information about weather was transmitted from continent to continent on a daily basis, using all types of waves from the shortest (18 m) to the longest (22,000 m). Two organizations were in charge of meteorology and radiotelegraphy: the International Meteorological Committee and the International Commission for Air Navigation. Experts decided that even the propositions to the 1920 conference (the so-called Washington project) were 'not suitable at the current stage of international meteorology,'[53] and further adjustments to the Radiotelegraph Regulations were therefore required.

The discussions exclusively concerned an international meteorological service. A special commission on meteorology decided that the weather reports

50 Lommers, *Europe – on Air: Interwar Projects for Radio Broadcasting*, 88.
51 Adrian Mackenzie, Wirelessness. Radical Empiricism in Network Cultures (Cambridge (MA), London: The MIT Press, 2010), 15.
52 Simon J. Potter, Broadcasting Empire: The BBC and the British World, 1922–1970 (Oxford: Oxford University Press, 2012), 49.
53 ITU Archives, "Documents de La Conférence Radiotélégraphique Internationale de Washington 1927. Tome I (Washington, 1927)" (Bern: Bureau International de l'Union télégraphique, 1928), 683.

could be divided into two categories: radio messages addressed to the general public and technical communication between specialists. It was noted that the latter type of message was not being transmitted in 1912 but was booming in the 1920s and represented the biggest change to have occurred in the domain of meteorology. If radiotelegraphy had been a promising yet supplementary communication channel in the 1910s, it became an absolute necessity for meteorologists in the late 1920s. This shift necessitated a high speed of transmission and adequate protection from interference. The convention decided that the meteorological messages and time signals should be transmitted using a fixed timetable, and other stations should remain silent, so as to avoid interfering with the transmissions. The International Meteorological Committee also organized the exchange of technical information from country to country between specialized meteorological centres.[54]

The weather reports made available to the general public, however, did not garner much attention in these international discussions. The 1912 regulations were perfectly adequate for the organization of such reports. The basic idea remained the same, namely, that the message should be clear, should be easy to pick up, and should be transmitted for a limited amount of time per day. This loss of interest in the public weather reports may have been provoked by the fact that they had become less relevant as an international service. Furthermore, the method used by the national services when transmitting weather reports, which involved paying particular attention to local particularities, could sometimes be much more readily understood.

This shift from international to national networks is also particularly evident in the scheduling of time signal broadcasts. Whereas they had previously been an international service, these signals were now transformed into purely national content. The timetables of the radio transmissions show that the international time signals were staggered with similar national programs. Moreover, national broadcasting stations might even develop their own idiosyncratic approach to transmitting the time signal. For instance, in 1924, the BBC began transmitting its iconic six-pip time signal as the Greenwich Time Signal.[55] Time signals and weather reports now reached the final listeners in the guise of a national product.

54 Ibid., 684–85.

55 J.A. McIlroy, "The History of the Greenwich Time Signal from 1924 to the Present Day," *Engineering Science and Education Journal* 2, no. 6 (1993): 281, https://doi.org/10.1049/esej:19930079.

Mapping global networks

Maps of radiotelegraphic networks represent an exceptionally rich source for understanding the state of radiotelegraphy and its politics. As radiotelegraphic stations proliferated, so too did radiotelegraphic maps. However, for a long time such maps were only produced as a courtesy by particular companies, institutions, or states, therefore representing only a fraction of the existing radiotelegraphic networks. For instance, the map of the Marconi Company (Figure 23), which was submitted as a proposal for developing the imperial chain after the war, represented only flows coming from the center of the Marconi Company in London, with no account being taken of any other networks.

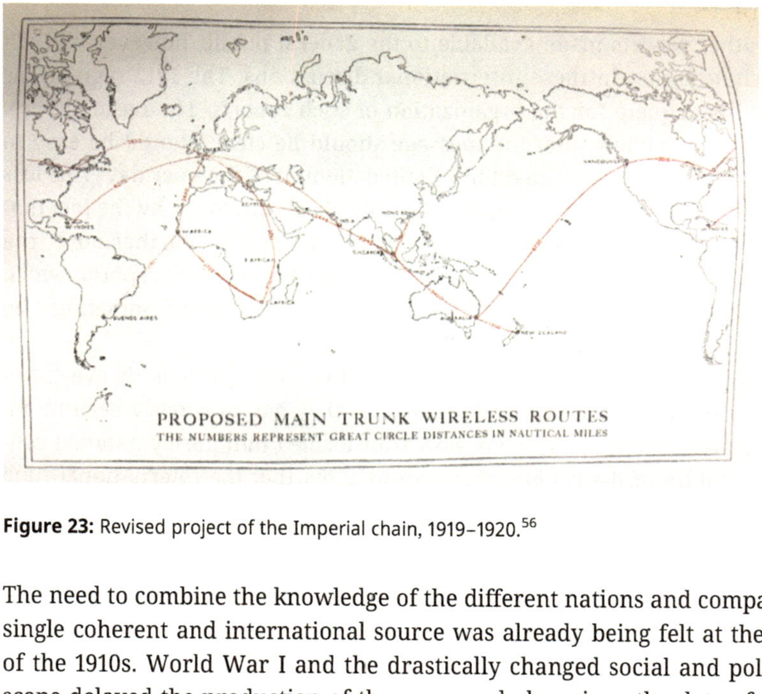

Figure 23: Revised project of the Imperial chain, 1919–1920.[56]

The need to combine the knowledge of the different nations and companies into a single coherent and international source was already being felt at the beginning of the 1910s. World War I and the drastically changed social and political landscape delayed the production of the maps and also, given the date of their issue, rendered them too controversial. Overall, the ITU produced three maps:

56 Marconi's Wireless Telegraph Co. Ltd. (1920). Proposal for a network of wireless communications to serve the needs of the whole British Empire. In SC MSS 143/8 (Printed Papers). IET Archives.

- An official map of radiotelegraphic stations on several large sheets of paper in 1922/23
- The same map in a second edition published in 1925/26
- A map of communication channels via radiotelegraphy between fixed points issued in 1927.

The ITU maps were some of the very first maps of radio networks to be made at the European level. The act of mapmaking was itself empowering; the actor that constructs, modifies and circulates a map has a unique capacity to put in dialogue a series of local knowledges about the world, and to provide one unified vision of it. Through maps, elite and powerful groups were able to promote a particular vision of the world to weaker groups.[57] A map never reflects reality but instead always reproduces it by simplifying, connecting, silencing, and shaping complex relationships. By designing a map, the ITU affirmed its privileged position as a superior actor able to standardize information supplied by the various state-members, to recombine it in its own favor and to distribute it as if it were an adequate representation of the international progress made.

The history of the production of these maps sheds some light on the transformation of radio. In 1911, the ITU began receiving requests to provide maps of radiotelegraphic stations,[58] however, although it published and regularly distributed a list of radio stations, the ITU did not as yet have any maps. This issue was put on the agenda at the 1912 London Conference, which resolved to satisfy the need for maps. The new convention thus obliged the ITU to produce and publish radio communication maps that also detailed their particular use for ships at sea. Throughout 1913, the ITU and its members exchanged letters, reports, lists, and national maps in order to collect all the requisite information regarding radiotelegraph stations. The different states and commercial companies sent their contributions in various formats: texts, charts, and sometimes using the ITU form. At the end of 1913, the ITU hired a Swiss cartographic company, Kümmerly and Frey, to draw the map. The first draft, which cost the ITU 2,300 Swiss francs, had been finished by the beginning of 1914,[59] however, because of the war and the continu-

57 Harley, J.B. "Maps, Knowledge, and Power." In The Iconography of Landscape: Essays on the Symbolic Representation, Design and Use of Past Environments, 277–313. Cambridge: Cambridge University Press, 1988.

58 e.g., ITU Archives, "D. 23. N. 12. Falett Co.," in *Registres de Correspondance: Radiotélégraphie*, 1911; ITU Archives, "D. 23. N. 1. Cie Franc Maritimes de t.s.f.," in *Registres de Correspondance: Radiotélégraphie* (1911).

59 ITU Archives, "D. 23. N. 6. Kümmerly and Frey," in *Registres de Correspondance: Radiotélégraphie* (1911).

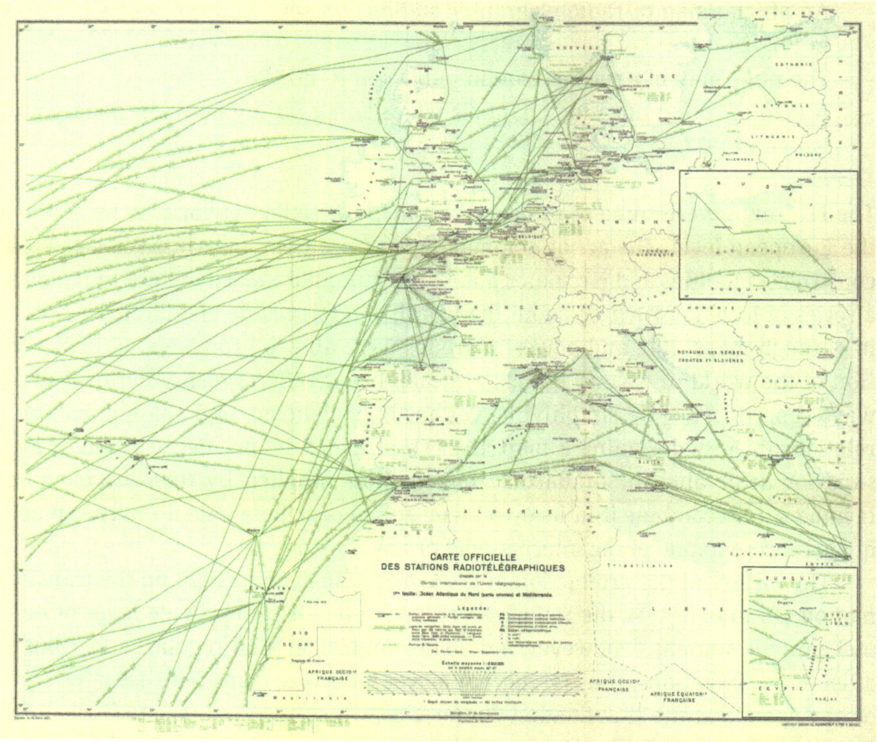

Figure 24: The ITU map of radiotelegraphic stations in Europe, 1922. The stations are depicted as dots and are hardly visible.[60]

ous changes to the political landscape, the publication of the maps was deferred. Several new drafts were made during this time to incorporate corrections, but none of them were distributed among the members. In the 1920s, those maps no longer reflected reality, as many radio stations had changed. As a result, supplementary data had to be collected, which took a great deal of time. Kümmerly and Frey had already been asked to make modifications to the existing map in 1919, but changes to the landscape of radio communication had been so great that the new data simply did not fit the existing template. The first two sheets of the new map were published in June 1922 (Figure 24), with the final two sheets to follow the next year.

60 ITU Archives, *Carte Officielle Des Stations Radiotétégraphiques, 1° Édition, 1922–1923* (1923).

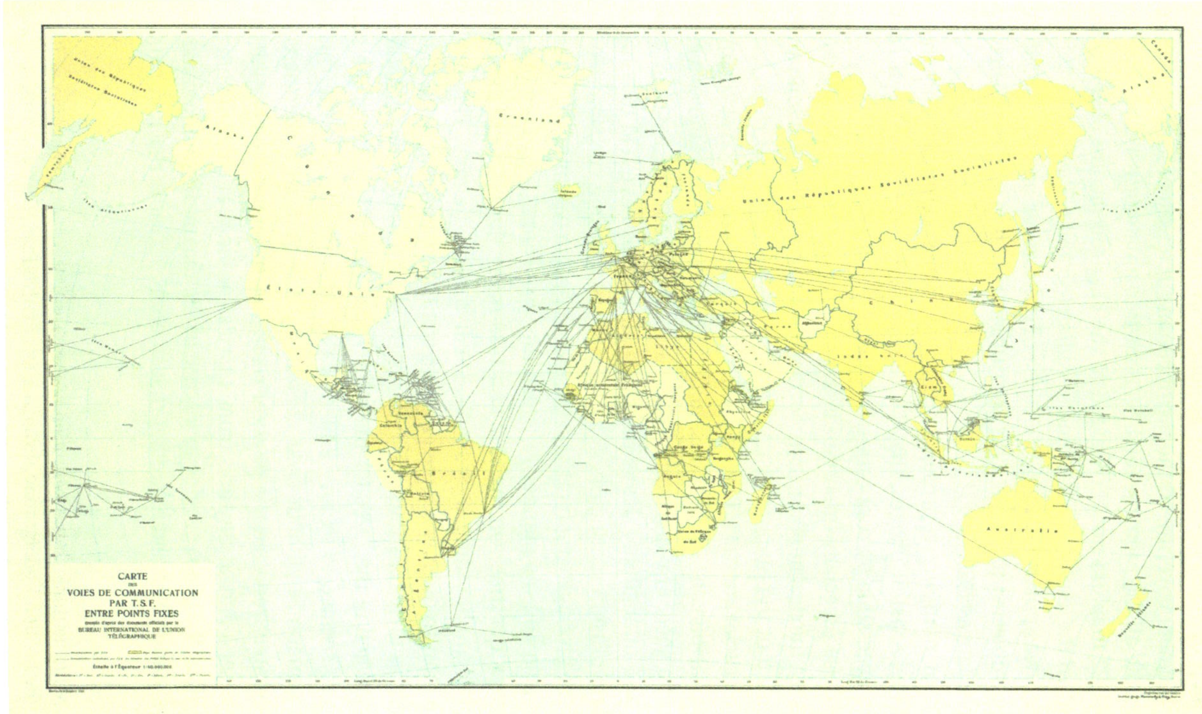

Figure 25: The ITU map of radiotelegraph channels, 1927.[61]

61 ITU Archives, *Carte Des Voies de Communcation Par TSF Entre Points Fixes* (1927).

The published map of 1922/1923 was accompanied by additional comments and came with a supplement covering all the modifications in a textual format. This inability to incorporate all the requisite modifications into an already drawn draft indicates that the 1922 map did not correspond to the changed picture of the world and indeed reflected the 'old' understanding of radiotelegraphy. The second version of this map, which was published in 1925/1926, had not however been drastically altered. After being subjected to a great deal of criticism, the map was completely redrawn, but only as late as 1927.

The process of mapping represents the social construction of knowledge about radio. In these visual images, the ITU renders the understanding of radio more simple and accessible by connecting it with the physical reality.

There were three especially important issues to be addressed by those drawing these maps, namely, the design of the map, social context, and political significance.

First, the design of the map, in its depiction of the invisibility of radio waves, was addressing a crucial question. Like any other visual material, such as the previously mentioned cartoons or illustrations, maps also had to show radio waves on the flat surface of a sheet of paper while simultaneously accounting for their absolute invisibility and intangibility. Many infrastructural maps have had to confront this same problem of representing phenomena that are invisible yet material, and disconnected to the given geographical space.[62] Maps of radio communication depict invisible phenomena and the manner in which they are connected to the physical and visible space in question. On the 1922 and 1925 maps, the ITU's choice was to depict the radiotelegraphic stations as simple, small dots on the map. This method was also the one most frequently adopted in the fragmentary national maps that the ITU members sent as their contributions in the 1910s, as well as by other companies. The draft maps created in the 1920s for the radio broadcasting services of the Marconi Company also seemed to feature a plethora of dots representing the various stations,[63] even sometimes without clear contours of the continents themselves.[64] This method accurately reflected the physical reality, as these radiotelegraphic stations were indeed seen only as tiny dots from a bird's eye view. The disadvantage of this method was that it did not provide any sense of the magnitude of the radio station. The dot of a station also had a number that stood for its radio reach in nautical

62 William Cartwright, "Rethinking the Definition of the Word "Map": An Evaluation of Beck's Representation of the London Underground through a Qualitative Expert Survey," *International Journal of Digital Earth* 8, no. 7 (2015): 522–37, https://doi.org/10.1080/17538947.2014.923942.

63 "Plan Showing Principal Stations of Proposed Press Association Wireless Scheme. Dwg No 16726" (University of Oxford, Bodleian Library, ms. Marconi 301, 1920).

64 Marconi Archive, "Wireless News Scheme. Europe [Hand-Drawn Map]" (University of Oxford, Bodleian Library, ms. Marconi 301, 1920).

miles, along with information as to whether the station was public, private, or used for any other type of correspondence. However, someone reading the map still had to calculate whether the station in question was available. Therefore, this map still required advanced technical skills and did not help most viewers understand the radiotelegraphic network.

The 1927 map represented a gigantic leap forward in understanding the invisibility of radio by virtue of presenting all radiotelegraphic connections as lines. In fact, the name of the map – *Map of communication channels via radiotelegraphy* – already shows that it was not a map of the stations but instead a map of the information flows, which was already more relevant for the 1920s (see Figure 25). That was an important shift in understanding radiotelegraphy: from a map of stations to one of communication channels. None of these maps, however, illustrated the radiation of the stations' signals with circles, as is commonly done today, even though this type of design was already gradually appearing as a shared understanding of radio's reach in the 1920s.

Second, these maps address the social context of the existence of radio. What is somewhat peculiar about the maps of 1922 and 1925 is the sheer number of lines of sea navigation that visually dominate the graphics. With the green lines that connect the ports, the maps depict most of the regular maritime routes around the globe, with details of their length and the duration of the transfer. This relation to navigation at sea had been particularly valuable in the context of the 1910s, and it was a product of pre-war discussions approved at the 1912 London Conference. Article 5 of the Service regulations annexed to the 1912 Convention read as follows: "The International Bureau shall draw up, publish, and revise periodically an official map showing the coast stations, their normal ranges, *the principal lines of navigation*, and the time normally taken by ships for the voyage between the various ports of call."[65]

Moreover, the particular attention given to the sea on the 1922 map is evident from its composition of four sheets that were each named after oceans: for instance, 'North Atlantic Ocean (eastern part) and the Mediterranean' instead of simply 'Europe,' which it featured. The ITU map implied that only ships following the routes depicted could benefit from radiotelegraphy. On the 1922 map, the radio was entirely bound to the sea, which corresponds to the vision of radio in the 1910s. The 1927 map, however, was no longer focused on the sea, instead emphasizing communication between land stations, which represents a crucial turn during the war period from the sea to the land (see Figure 26). In 1922, only coastal radio stations were shown on the map; the space inside the national borders was intentionally left completely blank or was used for the additional textual comments. The national

65 Italics added, ITU Archives, *International Radiotelegraph Convention (London, 1912)*, 188–89.

space was indeed 'silenced' on this map. The principal radio stations on the 1922/1923 map were all on the coast, while the 1927 map featured stations located in the capitals and principal cities of the different countries. Thus, the map of 1927 features Paris as the main center of communication routes, while on the 1922 map it was not present at all. These capitals represented the centres of communication, as they were the starting points of most of the lines of radiotelegraphy for the 1927 maps. Moreover, some of the lines were depicted with arrows going from European empires to their colonies, such as one from Paris to Dakar. This represents an important shift in understanding radio communication: from ship-to-shore to land-to-land.

Furthermore, the political valency of these maps was also quite different. As mentioned above, the 1927 map presented the national capitals, which for the first time demonstrated the importance of the nation in radio communication. The history of the map is linked to the triumph of the nation-state in the modern world, and a map, historically, has always depicted the world from a national perspective and shaped the idea of a national community. Even global and therefore seemingly 'neutral' maps intended for an international audience still have their own national preferences and emphases. Because of its international status, the ITU had to provide an internationally 'neutral' vision of the radio, but this general character was determined by the member states and was mostly more European, with the interests of particular countries predominating. They were a starting point for most of the lines of radiotelegraphy. Moreover, some of the lines were depicted with arrows, so as to indicate a one-way flow, from imperial capital to colony.

The first map had lacked this national bias, however, focusing rather on the sea as a transnational space. The nations were already taking on a more crucial role in the second edition of the map, produced in 1925, when one of the sheets was not named after an ocean but after the names of the countries covered, namely, Greenland, Iceland, Norway, the USSR, and Alaska. The 1927 map was not focused only on the sea but instead represented a combination of different national perspectives with Europe right in the center of the map. As Badenoch suggests, this shift could be interpreted in one of two ways: not only was Europe presented as the center of a uniform and interconnected communication network, but Europe was also 'where the network is.'[66] Furthermore, the presence of the US on the map created a vision of a global communication space, which the existence of radio stations on different sides of the country implied further, because it enabled radio waves to traverse and

66 Alexander Badenoch, "Myths of the European Network: Construction of Cohesion in Infrastructure Maps," in *Materializing Europe: Transnational Infrastructures and the Project of Europe* (Basingstoke, Hampshire and New York: Palgrave Macmillan, 2010), 47–77, https://doi.org/10.1057/9780230292314_4.

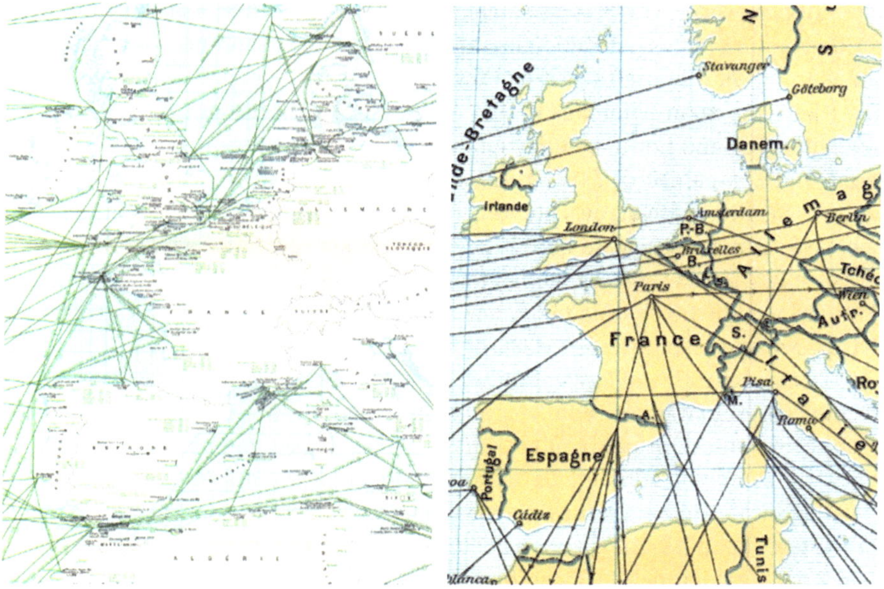

Figure 26: Fragments of the ITU maps from 1922/1923 and 1927 featuring changes in radio communication: from ship-to-shore to land-to-land.[67]

unite the entire globe. The US was also depicted as an important hub for connections with other parts of the world, such as Greenland or Mexico.

These attempts to collect the data and combine them on a map thus indicate that the ITU was bearing witness to the transformation occurring in radio. Initially it had depicted the sea as the only area of radio communication, eschewing the issue of radio as a national medium. The sea had been presented with all its shipping, and each and every ship could potentially call any shore station; it was a single, unique space for transnational communication. However, the publication of the second edition of the ITU map three years later signifies that there were modifications that had perforce to be added in order to reflect the new political and social landscape. The publication of an entirely new map in 1927 indicates an utterly different idea of radiotelegraphy: a global perspective, with all the continents fitted on to one sheet and battling over the information channels targeting the colonies.

67 ITU Archives, "Carte Officielle Des Stations Radiotétégraphiques, 1° Édition, 1922–1923" (1923); ITU Archives, "Carte Des Voies de Communication Par TSF Entre Points Fixes" (Geneva: Bureau International de l'Union Télégraphique, 1927).

Although the 1927 map was smaller and easier to read at a glance, it offered an illusion of a coherent radiotelegraphic space. This coherence was visible on the map but hardly useful, just like the first maps of highways and roads in the US, where the main point was to plot the physical presence of thoroughfares on the landscape and not to actually guide drivers.[68] The networks of different nations and companies co-existed in the two-dimensional scheme but were not united. The lines with arrows pointing towards the colonies implied an important message; that space was to be fought over and not to be shared. In this way, it drastically differed from the map of the newly established IBU that suggested that European countries share and organize the radio spectrum to avoid interference (also discussed on p. 163, Figure 22).

To conclude, the analysis of these maps demonstrates that the primary focus in the representation of radio communication on an international level shifted from communication between continents across the sea to Europe as the center of world communication. Radio became more centralized in the hands of European political capitals; on the following map, we can see a more schematic image of the world and less information about radio stations, which indicates that radio had finally gained political significance for European governments. It began to develop within national borders, proceeding from the capitals to the colonies, as a national rather than a transnational project.

Institutionalization of radio amateurism

During the 1920s, the position of the radio amateur had dramatically changed. On the one hand, the technical experiments with the technology continued, promoting the use of radio and leading to its proliferation within the home environment. On the other hand, as radio proliferated in private dwellings, it became more user-friendly and concealed the technical side of the technology. Radio underwent 'blackboxing,' in other words, the technical details were hidden and the focus was increasingly on user-friendliness, which could also be seen to reflect the success of wireless. As Bruno Latour put it, 'the more science and technology succeed, the more opaque and obscure they become.'[69]

---

68 Adam Plaiss, "Who Gets to Draw the Map? The Contentious Creation of the American Road/Map System, 1917–1926," *History and Technology* 28, no. 1 (2012): 7.

69 Bruno Latour, *Pandora's Hope: Essays on the Reality of Science Studies* (Cambridge, MA: Harvard University Press, 1999), 304.

The new international regulations governing radio broadcasting, discussed earlier, had massively contributed to this process. As Fickers has shown, the regulatory decisions of the 1920s, starting mostly with the IBU initiatives, unintentionally resulted in what he called the 'visualization of the hearing experience,' the radio dial, which offered the experience of 'surfing' on the radio waves, but without any of the fine-tuning that had been common in the 1910s. It displayed a list of specific stations available to the listener, offering 'an imagined ether voyage, where London, Paris, Oslo, or Hilversum were just one turn away from each other.'[70] Thus, devices throughout the 1920s acquired a comfortable design, which encouraged their use by all family members. This user-friendliness (featuring, among other things, a radio dial), transformed the radio receiver into a domesticated piece of furniture.[71]

With the development of the radio industry and the domestication of the technology, amateurs (radio hams) and general users of radio slowly drifted apart. This meant that not only did the number of radio users increase through the inclusion of less-experienced technology users, but radio amateurs were somehow segregated off. Considering how technically advanced and skilled radio amateurs were, their activity had by the same token to be managed, organized, and limited.

From a transnational point of view, the war had enhanced cross-border knowledge and technology exchange and practically levelled up the development of radio in Europe, bridging the gap between higher achievements and slow progress in the different countries. One notable example is that in 1918, UK amateurs expressed their delight at improvements in the radiotelegraphic equipment of the French installations because it rendered signals more regular and vigorous, and therefore, in in contrast to the pre-war years, more easily received from abroad.[72]

Communities of amateurs still flourished. Local clubs grew into regional associations, finally giving rise to nationwide organizations which were centrally organized; they collected information about hams, published call-books, and acted as a bureau for sharing contacts. They created a hierarchical structure that made it possible to control and manage previously scattered individuals. The London Wireless Club took other clubs under its wing, becoming in 1922 the Radio Society of Great Britain. That same year, Ireland similarly formed its own national association, the Radio Society of Ireland. In 1925, the first German Radio Technical Association was founded (Deutsche Funktechnische Verband, DFTV), which combined a large number of previously separate regional associations. The same year, the

70 Fickers, "Visibly Audible: The Radio Dial as Mediating Interface," 32.

71 Fickers, "Visibly Audible: The Radio Dial as Mediating Interface."

72 "Progress in France," *The Wireless World* (1918): 91.

Reseau des Emetteurs Français (REF), an association of radio amateurs, succeeded its predecessor, the Union des Sociétés de TSF de France. A year later, Portugal launched a Network of Portuguese Transmitters (Rede dos Emissores Portugueses) in 1926, while in 1927, the first national society of radio amateurs in Italy was formed, the Associazione Radiotecnica Italiana, followed by the Union of Swiss Short Wave Amateurs in Switzerland in 1929. The radio amateur community had become national in range and scope.

The institutionalization of the radio amateur community is also evident from the spread of the QSL-cards in the 1920s. As mentioned, radio amateurs exchanged information by means of the Q-codes. One of the codes was QSL, which also lent its name to the QSL-cards, working as follows: once one radio amateur had reached another, they would exchange some information about themselves using the codes. Any routine question, such as one about type of station, address, or name, had its equivalent in Q-code. Typically, radio amateurs would also ask each other a QSL, which means 'Can you acknowledge a receipt?' They would exchange postal addresses and send each other actual, physical postcards, to prove that the radio connection had indeed happened, which were known as QSL-cards. From the 1920s, these cards became more formalized and acquired a standardized format, which also indicates a degree of institutionalization of the community. The centralization of such collections, resulting in the creation of the national QSL-bureaus, is also one of the consequences of such change.[73]

Amateurs introducing broadcasting

Ham radio organization set itself at least two goals. First, the national associations provided a degree of control; they issued licenses and obliged amateurs to take specific exams and to abide by the relevant national and international legislation. The ambition was to render the radio spectrum much clearer and thereby avoid any disturbances. Second, through these associations, governments could elicit and collect certain information about the radio. In doing so, they followed the path taken during the war, when amateurs were engaged in developing and explaining the uses of different devices. One of the more notable cases is that of France, where amateurs were encouraged to send in their feedback on experiments with radio broadcasting transmissions from the Eiffel Tower. After the end of the war, the mil-

73 Maria Rikitianskaia, "Listening to "Concert of Europe": Pioneering Radio Amateurs during World War I," in *World War I: Media, Entertainments & Popular Culture*, ed. Chris Hart (2018), 122–42.

itary authorities executed several tests on the Eiffel Tower regarding the use of short waves, with the amateurs invited to collaborate by submitting the results of their observations along with any notes they may have taken about the receiving devices.[74]

Radio amateurs and radio engineers constituted the very first generation of radio professionals, with some of them even responsible for launching radio broadcasting stations. For instance, French radio engineer Maurice Vinot decided in 1922 to embark upon his own project, which he dubbed Radio-Gazette, later known as Paris-information-sans-fils (PISF). This project originated in Maurice Vinot's keenness to receive meteorological information from the Eiffel Tower, using a renovated marine radio receiver given to him by his friend radio engineer Elie Podliasky. The PISF provided radio devices for domestic use and a complete service of information and news from all over the world, from political, domestic, and foreign affairs to theatrical chronicles. The service was supposed to run without interruption from 11 a.m. to 11 p.m., and the most highly qualified men of science, along with practitioners from the world of the arts and even comedians, were invited. It was designed to cover around 30 kilometres, meaning the zone of Paris and the surrounding area.[75] In 1922, Vinot also submitted a request that he be authorized to install a radio receiver in his car so as to be able to catch radio broadcasting on the move, thus anticipating the later development of radio.[76]

Broadcasting soon became seen as a propaganda tool, while radio enthusiasts for their part became an important human resource for the spreading of radio. One of the arguments of Maurice Vinot for convincing the government, for instance, was that a radio-magazine would be very easy to control by the administration and could be used as an efficient propaganda tool.[77] The Bolsheviks likewise saw radio as a nation-building instrument; a radio station was erected in every important Russian city, and they even introduced a new political programme called 'radiofication.'[78] People were encouraged to build radio stations, establish radio societies, and share their technical knowledge.

74 "Courrier Des Amateurs de T.S.F.," *Le Petit Parisien*, May 11, 1924.

75 Rebecca P. Scales, *Radio and the Politics of Sound in Interwar France, 1921–1939* (Cambridge: Cambridge University Press, 2016), 35–36.

76 Duval, *Histoire de La Radio En France*, 37.

77 Ibid., 35–36.

78 Stephen Lovell, *Russia in the Microphone Age. A History of Soviet Radio, 1919–1970* (Oxford: Oxford University Press, 2015); Maria Rikitianskaia, "How Children Learned to Listen: The Formation of Radio Clubs in the Soviet Union [Kak Detej Uchili Slushat'(Sja): Stanovlenie Radiokruzhkov v Sovetskom Sojuze] (in Russian)," *Logos* 27, no. 5 (2017): 141–62.

This usefulness of the amateur was even more pronounced in the Soviet Union. The Bolsheviks were all too keen to use radio as a mass media device, and indeed to develop it to its fullest extent, however, shortages of raw materials and the patchy development of the radio industry presented them with particular difficulties. Amateurs were therefore entrusted with the task of delivering radio into people's homes. By September 1918, two new radio journals had already been launched, each with a different target audience; the more academic *Telegraphy and Telephone without Wires* was reminiscent of the scholarly literature of the Russian Empire and targeted the academic and advanced amateur, while the second journal, *Radio Technician*, was something of a novelty, addressing as it did a vast audience. In the latter publication, complicated aspects of radio and wireless telegraphy were explained in a popular manner, applicable to everyday life tasks. No radio amateur society had adopted such a format before; indeed, all other journals consisted of formulas, scientific book reviews, and observations regarding recent events. In other words, *Radio Technician* became the first Russian periodical to provide clear instructions on how to build a radio device and thereby become a full member of the radio amateurs' society. The radio amateur community grew steadily as new books on wireless appeared and the first clubs were formed. In December 1918, the First Radiotelegraphic Congress was inaugurated in Petrograd. Later, in 1923, the Council of People's Commissars promulgated two decrees, which were extremely important for the development of broadcasting in the country at large. The Decree on Special Purpose Radio Stations allowed all state, professional, party, and public institutions to construct and operate radio reception stations. A year later, a second decree was issued by virtue of which all citizens of the USSR were allowed to possess radio receivers; they could be bought in a store or designed independently. This decree contributed to the rise of the radio amateur, it having been no easy matter to obtain a license in the tsarist period. In the years after the revolution, there had been almost no radio amateurs at all. In 1923, the first national association of amateurs was formed: the Society of Friends of Radio. The particular task of this organization was introducing radio to people not as a point-to-point medium but as a broadcasting mass medium. The relevant specialist literature and the more general press coverage had taught readers to use radio not for internal communication but instead for 'listening to Moscow.'[79]

In particular, amateurs were encouraged to involve children in spreading radio, just as had been done during the war. The policy targeted children because they were curious about physical experiments and showed great interest in radio.

79 "Radiofication in the USSR [Radiofikazia SSSR]," *Radio Vsem* 7 (1927): 145.

Children would usually lose no time in becoming more competent than their teachers:

> Everyone . . . undoubtedly notices how big is the gap between schoolchildren and teachers in the matter of radio construction. The young generation, both in clubs and alone . . . seeks to learn and build a radio receiver . . . They are using previously never used words and expressions from the field of radio engineering . . . like 'receiver', 'detector', 'hung up an antenna!', 'it does not have a condenser', 'where would you go with your capacity', etc. And it covered literally everyone, from pioneers to adults.[80]

Overall, amateurs in Russia helped to organize and promote an active radio amateur community that did much to spread radio across the country. Moreover, this radio was exclusively designed to obtain information from the centre. These radio experiments did not serve to establish international contacts; indeed, the primary aim was not to establish any such contact, and most communication between radio amateur clubs was by letter or by way of publications in journals. Amateurs only helped to make broadcasting more available.

Renegotiating gender roles

The radio was seen for the most part as entertainment for men. Women were in this regard represented as wives who simply bore witness to the passions of their husbands, seen with a cartoon from 1923 which presented the wife of a British radio amateur as if she had lost her husband to his hobby, with the subtitle dubbing her a 'wireless widow' (Figure 27). The picture is also symbolic inasmuch as it depicts a husband listening by himself to a broadcast; he wears headphones and sits comfortably in his chair at a distance from the device, fully absorbed in his individual listening experience. This depiction of his practice of operating the device differs markedly from those current in the 1910s. In 1911, for example, a drawing showed a passionate amateur actively adjusting the device, with the public apparently deeply involved (see more on p. 99); in contrast, the 1923 illustration captures a passive listening experience. In the 1910s, the signal needed to be 'caught,' while in the 1920s, people were already comfortable passively listening to a continuous flow of information.

Nonetheless, the number of women who operated radio devices had significantly increased after World War I. The presence of women could be explained by the war, as it had mobilized all possible resources, including women, for radiotele-

80 "The Enlightener. It Is Necessary to Help School [Prosveshhenec. Nado Pomoch' Shkole]," *Bjulleten' Radio* 1 (1925): 20–21.

Figure 27: 'The Wireless Widow,' 1923.[81]

graph work. The US magazine *QST* evinced surprise when reporting this fact: 'One of the last things we expected to take place was the advent of Woman in Amateur Radio.'[82]

Even though a fair number of physicists, electrical engineers, and chemical engineers were female (Marie Sklodowska Curie, for example), the presence of women in the amateur radio community was still somewhat shocking.[83] The ability of a woman to take over such work seemed startling: 'When it comes to sitting beside a rotary gap which is snorting blue fire and emitting ear splitting shrieks,

81 Lendon. The Wireless Widow. *The Sketch* (1923, June 6): 467.
82 "The Radio Ladies", *QST*, March (1921): 30.
83 "The Radio Ladies (Concluded)," *QST* (March, 1921): 64.

we cannot refrain from wondering.'[84] The American amateurs noted that the presence of women affected the language of communication and transformed jargon in particular; the term 'old man' (OM) had typically been used for an amateur, but the number of female amateurs had increased to the point that the name was no longer suitable.[85]

In contrast to the US, women in Europe had proved themselves to be useful during the war and were no longer regarded with skepticism. While the Americans were still wondering just how to address a woman, the radio broadcasting stations in European countries, which were growing rapidly, showed no hesitation in employing women to fill the different positions. Women were thus all too able to pursue successful careers in the radio sector in the 1920s, when radio stations gradually moved from ships to land and focused on managing national rather than international communication. The stations, developed in the 1920s, also expressed their modernity by adopting a policy of gender equality. In particular, the BBC was known for its non-gendered grading system with equal pay and equal promotional chances for men and women alike.[86]

Thus, throughout the 1920s radio witnessed a rapid increase in the number of its users: professional technical experts, passionate amateurs, and knowledgeable users had grown not only in numbers but also in diversity.

International amateur service

The spread of amateur communities from a local to a national level reached its peak in 1925 when the International Amateur Radio Union (IARU) was established.[87] Twenty-three countries were combined within this organization: Argentina, Austria, Belgium, Brazil, Canada, Czechoslovakia, Denmark, France, Finland, Germany, Great Britain, Hungary, Italy, Japan, Luxemburg, Netherlands, Newfoundland, Poland, Spain, Sweden, Switzerland, Uruguay, and the US.

The whole idea of establishing such an organization had emerged from the meeting in Paris in 1924 between delegates of national radio amateur societies

84 "The Radio Ladies."

85 "The Ladies Are Coming," *QST* (August, 1917): 19.

86 Kate Murphy, *Behind the Wireless: A History of Early Women at the BBC* (Basingstoke and New York: Palgrave Macmillan UK, 2016).

87 "First International Amateur Congress. A Record of a Strenuous Week," 1925.

from the US, Belgium, Spain, France, Great Britain, Luxemburg, Italy, and Switzerland. Denmark was unable to send a delegate but provided a letter of support.[88]

The original constitution of the Union provided radio hobbyists with individual memberships because national societies did not as yet exist in all of the countries involved; members were sometimes issued with a membership card (Figure 28). In 1928, however, the model changed in favour of a federative one. Individual memberships were eliminated, and the IARU focused on supporting the national societies, which it still does today. It was decided that each country around the globe could form a so-called national section, so long as it boasted a minimum of 25 members. The IARU thus came to occupy the summit of the hierarchical radio amateur community, however, the supreme authorities were still the national governments. The IARU was a private international organization with semi-public functions, which would represent radio amateurs at all international radio conferences from 1927 onwards.

The new organizational structure of the radio amateur community around the globe also indicated another important fact, a shift in the geography of power, as the center of this international control had moved from Europe to the other side of the Atlantic. The activities of the IARU in the 1920s were first reported in the pages of the American radio amateur journal *QST* and then spread to other countries.

INTERNATIONAL AMATEUR RADIO UNION
Headquarters: HARTFORD, CONN., U. S. A.

This is to certify that

IS A MEMBER OF THE INTERNATIONAL AMATEUR RADIO UNION AND IS ENTITLED TO ALL THE PRIVILEGES OF MEMBERSHIP UNTIL APR 2 4 1926

INTERNATIONAL SECRETARY-TREASURER

Figure 28: The membership card issued by the International Amateur Radio Union,1925.[89]

88 "An International Amateur Wireless Congress for 1925," *The Wireless World and Radio Review* (April 1924): 137.
89 "International Amateur Radio Union", *The Wireless World* (July 1925): 46.

Even today the ARRL, the national association for amateur radio in the United States, continues to serve as the International Secretary of the IARU.

Arguing at the Washington Conference for the need to have various different radio services, the delegates of 74 administrations asked for frequencies that greatly exceeded the number of useful frequencies available. Compromises therefore had to be arrived at.[90] By the time the Washington Conference met in 1927, issues concerning radio had become bewilderingly complex; the object of regulation – the radio – had been transformed beyond recognition both from a technological and a political point of view. The number of amateurs had also grown considerably, leading inevitably to interference with all other radio services. A new international agreement was necessary to put amateur experiments on a leash.

The Washington Conference would turn out to be one of the very few meetings in the international landscape where amateurs were offered the opportunity to speak up, even though they did not have an official vote. The IARU representatives were present at all discussions and participated in the conversations, but the amateurs also served as advisers to national delegations.

For the first time, the Washington General Regulations also discussed the role of an amateur in depth, and amateurs perceived the recognition of amateur radio as a 'great victory.'[91] The new convention had also defined the use of radio by private individuals: an 'amateur' was a duly authorized person interested in radioelectric practice with a purely personal aim and without pecuniary interest.[92]

In this definition, particular stress is placed on the words 'duly authorized,' which stands for the governmental authorization of amateur activity (i.e. an official license and agreement to obey international and national radio regulations). At that date, radio amateurs were defined by their institutional relations; they held licenses, had appropriate call signs, were listed in public call books, were members of an amateur union and commonly subscribed to one of the national radio amateurs' journals. This institutional perspective on radio amateurs' activity often dominated radio amateurs' handbooks and journals, giving new communities the opportunity to celebrate their birthdays. Moreover, it also assigned a particular character to the messages of amateurs; they had to be conducted in 'plain language' and be restricted to information about experiments and to re-

90 Peter B. Schroeder, "The Radio Amateur in International Legislation and Administration," *The American Journal of International Law* 48, no. 3 (1954): 423.

91 Kenneth B. Warner, "The Amateur and the International Radiotelegraph Conference," *QST* January (1928): 15.

92 ITU Archives, *International Radiotelegraph Convention of Washington, 1927 and General and Supplementary Regulations (Washington, 1927).*

marks of a personal character.[93] It gave the highest authority to the national level and forbad cross-border communication if it was not permitted in one of the countries in question.

As at the previous conferences, participating countries circulated copies of their proposals for the discussion to be distributed in advance by the ITU to all members. These proposals reflected each particular country's vision as to what radio might ideally come to be, with the most striking difference the manner in which the role of amateurs was envisioned. The proposals of Great Britain, Germany, Switzerland, Hungary, Japan, and the Netherlands involved the actual elimination of amateur radio as an international service. Only the US proposed quite a generous frequency band for amateurs to use.[94]

The 1927 Conference and the presence of the IARU members at the discussions demonstrated the big difference in the attitude towards amateurs evident in Europe and in North America. In the words of the IARU Secretary, the European powers did not support ham radio at all and even directed diverse and 'fallacious' arguments against amateurs. Moreover, the Europeans perceived the US government to be 'little short of insane in having encouraged amateur radio.'[95] Most of the arguments were motivated by a general fear of uncontrolled experimentation. The European nations were strongly supportive of a state monopoly on communications, which led to the general repudiation of contributions made by private individuals and commercial companies. In contrast, the army and navy officers of the US delegation claimed that amateurs could complement the armed forces, but this argument was given short shrift, as an amateur in Europe was viewed as a person who might undermine the security of the state by fostering the revolution and spreading 'red propaganda.'[96] In a technical sense, this fear was more than justified considering the crucial role of radio in the Russian Revolution of October 1917.

A special subcommittee was formed to discuss amateur concerns. Mr E.N. Shaughnessy from the British General Post Office presided over this debate, and the subcommittee also included President of the IARU Hiram Percy Maxim and Secretary Kenneth B. Warner. Most delegates supported the British view, namely, that the use of radio by amateurs ought to be restricted, however, this position shifted after the IARU representatives met informally with members of the various national delegations. Those informal meetings were not covered by the official conference

93 Ibid.

94 ITU Archives, "Documents de La Conférence Radiotélégraphique Internationale de Washington 1927. Tome I (Washington, 1927)," 683.

95 Kenneth B. Warner, "Editorial. With the International Radiotelegraph Conference Safely behind Us," *QST* (1928), 11.

96 Warner, "Editorial. With the International Radiotelegraph Conference Safely behind Us."

documents but were reported in terms of 'the afternoon tea-cups or in hotel rooms at night' in the IARU Secretary's summary of the deliberations.[97]

The final decision on amateurs supported the American vision. The frequencies were allocated following the pattern suggested by the US: five bands of frequencies in the short-wave region. This pattern is known as 'Hoover bands' and was named after Herbert Hoover, who had first suggested this idea at the US National Radio Conference in Washington in 1924.[98] Not coincidentally, Hoover was also the President of the 1927 International Conference.

A compromise in all these conflicts was reached by allowing each nation to choose to permit or prohibit amateurs, as it desired. It was also decided that the international amateur message traffic would be forbidden if one of the countries had given notice of its opposition to this exchange. Overall, this meant that the Washington Conference in 1927 recognized amateur radio as an international service. Controversially, message handling within national borders remained dependent on the decisions of the individual country concerned. The French delegation, for instance, strongly advised against the prohibition of third-party traffic. Therefore, according to the Washington Conference, an amateur could much more easily communicate with foreign colleagues than within his or her own borders.

Conclusions

Radio did not undergo a complete transformation immediately after World War I, but rather a lengthy and complicated overhaul lasting throughout the 1920s. The first international meetings on the matter of radiotelegraphy continued to be characterized by the attempt to restore global networks, creating the World Wide Wireless. There were also discussions to merge the electrical and radiotelegraph regulations in an attempt to create an intermedial network of electrical communications, however, this vision was challenged by the newly formed IBU, which underscored the importance of radio as a national broadcasting tool. The 1927 Washington Radiotelegraph Conference, which was organized by the ITU, defined radio communication and radio broadcasting for the first time ever and also allocated frequency bands for most of the possible uses of radio. The period of 'interpretative flexibility' of radio technology had ended, once again revealing the importance of radio as a national communication tool.

97 Warner, "The Amateur and the International Radiotelegraph Conference."
98 Schroeder, "The Radio Amateur in International Legislation and Administration," 422.

The radio industry changed significantly in the 1920s, opening up new possibilities for different companies. Radio broadcasting stations began spreading around Europe which were also all too willing to hire women, who had proved their usefulness during the war. Transnational networks of radiotelegraphy were no longer relevant because radio communication had acquired a more markedly national, rather than a transnational, character. This particular transition is perceptible on international maps of radiotelegraphy; the first ones had depicted radiotelegraph stations in relation to the navigation lines in the sea, so that radio communication could be used by anyone in this shared space, while the last one paid more attention to the role of nations and emphasized capital cities as the centres of communication routes.

Starting in the 1920s, radio amateurism also became concentrated around national societies and national radio stations. Some governments even continued actively exploiting amateurs in the development of the radio industry after World War I. The formation of an international union for radio amateurs may be said to exemplify such a national organization; it emphasized a centralized structure, regulated national societies, and all but eliminated the presence of any scattered individuals.

Chapter 5
The global wireless: Past and present

The development of the wireless telegraph in the 1910s was oriented on global communication, as the rapid spread of wireless networks around the globe promised to provide instant transnational communication even on the move. The radiotelegraph was seen to connect the world through wireless waves, contributing to a global communication space. It was supposed to work for the common good of the different nations, as they built coherent media infrastructures and developed collaborative projects. This global connectivity by wireless was so appealing that, in 1919, the radiotelegraph networks were called the World Wide Wireless, an expression strikingly similar to the contemporary World Wide Web.[1]

However, despite the bright prospects for global communication, plans to build this ambitious global wireless network have never been implemented. This utopian vision of wireless would never be realized, as the idea of a worldwide communication network did not survive in the 1920s world of rising nationalism. Notwithstanding the huge potential of global communication, the development of the wireless shifted towards nationally-driven projects after the World War I. The war had brought to the media environment a new idea of radio as a national medium that united the people of a particular country, which thus was nationally reorganized. This idea ousted the pre-war concept of radiotelegraphy as a transnational medium.

This understanding of wireless history in the early twentieth century sheds light on the development of radio broadcasting on one hand and wireless communication on the other. This chapter summarizes the most crucial findings of this research and discusses what they mean for media history and science and technology studies.

Wireless history as a progenitor of radio broadcasting

The history of wireless during the 1910s and 1920s serves as a perfect illustration of the interpretative flexibility of the new technology. When a new technology is developed, it will encounter a myriad different interpretations of its perfect use. As a confirmation of this rule, the wireless likewise had numerous different future paths for development, which is why the very definition of wireless in the

1 "World Wide Wireless."

Open Access. © 2024 the author(s), published by De Gruyter. (cc) BY-NC-ND This work is licensed under the Creative Commons Attribution-NonCommercial-NoDerivatives 4.0 International License.
https://doi.org/10.1515/9783111202327-006

1900s and 1910s is highly problematic. It is both a point-to-point and a one-to-many medium, both for private and public correspondence, Morse code and voice transmission. In fact, I have referred throughout this book to these sets of opposing features, in order to demonstrate how the technology swung towards certain interpretations, but simultaneously always also had an alternative use. Even when the predominant meaning of wireless was to offer a point-to-point communication, there were many applications of it as a broadcasting medium capable of reaching many people at once.

When a technology is open to different interpretations and is thus inherently flexible, it is reasonable to infer that its development has been shaped by various different forces. If we follow the STS approach, the dominant applications of a technology can be shown to depend on the respective powers of the relevant social groups. Those groups entertain certain visions for the technology's future and compete until the technology is properly set, which is usually referred to as a "closure of interpretative flexibility." The case of wireless telegraphy demonstrates that, interestingly, in its early history there were at least two closures that brought about radically different understandings of wireless: one in the early 1910s and another in the late 1920s.

The first "closure" could be attributed to the 1912 Titanic tragedy, and to the decisions subsequently taken for the development of the technology at the 1912 ITU conference. Pre-war radiotelegraphy was mainly seen as a product of transnational communication, and the early 1910s was the period when this predominant meaning of the technology emerged and provoked further stabilization of the technology. The 1912 Titanic disaster had revealed the importance of radio as an essential instrument for distant and transnational communication, and the benefits of radio as a technology for saving lives and bringing peace to the world drove most of the decisions taken at the ensuing international conference of 1912. Radio regulations, along with a number of new projects, such as synchronization over the radio with international time, presupposed that radio was a genuinely transnational technology. It was seen as an inherent part of the sea, *mare liberum*, something shared by many nations and therefore by its very nature transnational. Even the broadcasting experiments of the 1910s addressed a transnational audience, since they made no distinction between nations. The introduction of commonly understandable abbreviations and codes also mitigated national differences. Amateurs, in particular, mediatized and metabolized this new technology by involving their neighborhoods in the intriguing process of catching time signals or even creating new user-friendly devices. In sum, radiotelegraphy before World War I was thought to be a technology that was destined to spread across the world, transferring information swiftly and efficiently. This vision of the future of wireless could be seen as a closure of interpretative flexibility; after a decade of motley experiments, a clear understanding of the need for global wireless connectivity had finally emerged by the early 1910s.

However, the trajectory of wireless development was dramatically challenged with the outbreak of the war. During the war, the capacity of wireless to transcend borders was also hugely useful to government officials and the military alike, which highlighted a completely different aspect of radio as a military advantage and a means of national defence. The messages that were transmitted could travel across whole continents and be picked up on the other front. Due to the intensive development of different methods of cipher and coding, the use of radiotelegraphy was restricted to privileged nations or individuals. Most European governments had drawn up regulations for the use of radio waves, their aim being to establish "wireless silence," with the international climate too fragile to be threatened by experiments with wireless technology that might hinder governments' employment of wireless telegraphy and telephony for national defence. Furthermore, the war led people, companies and technologies to change their "nationality" and move across borders, which raised more concern than enthusiasm. In other words, World War I served to undo the apparent closure of interpretative flexibility as far as wireless was concerned, intensified change in the social groups affected, and allowed actors to consider a range of other options and scenarios regarding the future of radio technology.

The 1920s witnessed lengthy debates regarding the use of wireless and radio, as the transformation of radio during World War I had raised various controversial issues as to its potential uses. At least two contrasting visions for this technology were in play: one of global connectivity, the other of the national good. This competition between different visions occurred because the relevant social groups had changed over the course of the war. In particular, the geography of power had shifted from a Eurocentric vision to a more globally contested picture, because the US and even distant colonies had acquired stronger positions in the international arenas. The strategic use and management of radiotelegraphy over the war, along with the intense development of new technologies, had brought about permanent changes in the overall understanding of wireless. The admission that radio was of national importance had had a significant impact on its development in the 1920s, when it became a national service mainly directed at the citizens of the country where it was being transmitted. Over the course of the war, radio became a national responsibility, and, rather than being mainly focused on distant communication, the newly developed devices and technologies targeted predominantly national audiences instead. Thus, with the launch of public and private radio broadcasting stations, international collaborations gradually gave way to services that were of national benefit, and discussions about possible international projects were replaced by the drafting of regulations serving to divide national spheres of interest. The technology was reinterpreted, with a second closure of interpretative flexibility occurring in the late 1920s. The war had served to suppress the transnational meaning of the technology, on account of the national centralization of power that had taken place, and radio re-emerged in peacetime as a purely national communication instrument.

Therefore, because of the war, radiotelegraphy once again entered a period of interpretative flexibility, which later allowed radio broadcasting to be born. In the terminology of the 'double-birth' model, this process could be seen as a triple birth. Many scholars, including the actual founders of the model,[2] have acknowledged that throughout history media have in fact had multiple births, as they are always in the process of being reconfigured. This was the case with radiotelegraphy.

The concept of relevant social groups account for these drastic changes in wireless development and its gradual transformation into what could be termed nationally-driven radio broadcasting. The technology did not transform and experienced certain closures of interpretative flexibility on its own, but the relevant social groups involved in its social construction had changed over the course of war. The configuration of the different actors involved had altered; people had been mobilized, companies restructured or bought, and technologies transferred across national borders. New actors emerged; extra–European countries came to occupy a more prominent place in the international arena and even women, who had previously been marginalized, contributed to the socio-technological change under way.

The change in the relevant social groups happened in several arenas simultaneously. From a global perspective, the ITU actions became less pronounced. The period of European radio silence during World War I had forced the ITU to extend its focus from the European political space to other parts of the world. The actors involved in the ITU network became aware of the significant achievements and local characteristics of radio development in various countries, which had helped to establish the role of the radiotelegraph as a valuable replacement for disrupted telegraph cables. However, after the war, the ITU's reputation suffered, in the light of its failure to accumulate the necessary information about the recent changes and provide data quickly enough. Instead, other international meetings and the newly formed IBU were deciding the future of radio. In 1927, the ITU managed to regain its position in international techno-diplomacy through the conference in Washington and by virtue of its well-established global connections, its general approach to regulating all radio communications, and its proposals to incorporate the radiotelegraph into other telecommunication systems.

2 A Gaudreault and P. Marion, "Measuring the "Double Birth" Model against the Digital Age," *Early Popular Visual Culture* 11 (2013): 158–77.

Throughout this period, the ITU also bore witness to changes in the geopolitical and therefore the geographical context; colonies were coming to the fore, and there were suggestions that English be adopted as the official language for the international meetings. Concerns such as these also reflected the change in radio from a transnational network for sea communication to national and international services that both competed for the spectrum.

The main actors in the radio industry had also changed. In the early 1910s, corporations envisioned the radiotelegraph as a group of transnational networks and aimed to extend their business to as many countries as possible. During the war, only the news agencies continued working without any disruption of their infrastructures. This practice influenced the creation of different radio broadcasting stations in the 1920s and therefore confined the service to a more narrowly national scope.

The technical experts also gradually ceased to be the principal target audience for wireless. Originally radiotelegraphy had represented an important tool for technical experts only, however, with the spread of time signals and weather reports, this technical information could no longer continue to be the exclusive preserve of experts and became more readily available to ordinary users. The creation of national services in the 1920s reflected the recommendations made in these reports, but these services were designed for ordinary people and did not have anything in common with the technical information of the 1910s.

Furthermore, the radio amateur community likewise overcame challenges arising out of the war. Contrary to the commonly held assumption that radio amateurism began in the 1920s, my own historical analysis has demonstrated the existence of small amateur communities and individual experimenters throughout Europe in the 1910s. Even though they were not institutionalized, radio amateurs had already consolidated into small local societies and were attempting to construct devices before the war. From 1912 to 1927, they functioned as mediators between international organizations, governments, and their local communities. They followed the international agenda on radio regulations and exchanged news about inventions and devices in journals and through private correspondence. These amateurs also applied their knowledge of radiotelegraphy to their own particular fields of expertise, such as meteorology, seismology, medicine and philosophy, and introduced wireless communication to their colleagues. With the outbreak of war, their activities were challenged, since wireless technology had become an instrument of national power, and consequently something not to be entrusted to private individuals. However, even though governments imposed restrictions and bans on radiotelegraphy, amateurs were still involved in its development, either through their participation in military operations or in the guise of illegal domestic experiments. Therefore, in the 1910s, their activities had a more transnational scope, whereas in the 1920s these scattered individuals formed local clubs and then regional associations and national organizations. World War I did not provoke the birth of professional and amateur radio but rather, with the institutionalization, nationalization, and extensive development of radio broadcasting, hastened its segregation into various different fields.

Overall, these findings contribute to a new understanding of the history of early wireless and radio, by dint of adopting a transnational point of view. The trajectory of radio development is frequently described as moving from local to global, with Guy Starkey noting that trends in radio history are often characterized as 'influences behind and consequences of a paradigm known as "globalisation"'.[3] Such perspectives construe the history of radio as having its origins in the 1920s, while this monograph demonstrates the reverse process – how radio changed from being a transnational to a national medium.

Furthermore, a separate note should be taken of the advent of broadcasting and the use of wireless as a one-to-many medium. Wireless telegraphy could be seen as a precondition for the development of radio broadcasting, as it was indeed used to disseminate information around the globe. Contrary to the prevailing consensus in the current scientific literature, which dates the advent of broadcasting to the 1920s, this book has demonstrated that the origins of the phenomenon lie in the 1910s, preceding World War I. The transmission of time signals, weather reports, and news to ships rested to a great extent on the idea of spreading information simultaneously to various different recipients. Indeed, these transmissions were intended to be heard by the largest possible audience and were regularly scheduled to avoid interference with other transmissions. Radio operators on board ships were obliged to listen to these programmes, which therefore changed them from being senders of the messages to being listeners, and a synchronized audience too. The transmissions were scheduled at the same time every day, representing therefore what Crisell calls 'fixed point' scheduling,[4] as well as the very first 'programmes' on the radio. The organization of sea navigation and the introduction of codes, such as the SOS message, were designed to allow the information to reach a wider audience. The experiments undertaken by radio amateurs also proved their ability to spread signals and create synchronized audiences. Furthermore, even though these broadcasting services were mainly based on Morse Code transmissions, which complicated the use of technology for non-experts, voice and speech transmission had already been

---

3 Guy Starkey, *Local Radio, Going Global* (Basingstoke and New York: Palgrave Macmillan, 2011), xii.

4 *An Introductory History of British Broadcasting*, 49.

invented in the 1910s to facilitate the use of the radio. Many experimental projects before the war were already using the newest inventions to transmit music and the latest news with sound, either established ones, as in the case of the Belgian concerts, or smaller ones, such as the experiments conducted by a Portuguese amateur. Broadcasting was therefore already developed and adopted as a regular media practice even before World War I, with its organization and schedules set up to facilitate radio reception and target its audience directly. The advent of regular radio broadcasting stations in the 1920s should be seen not as a revolutionary process but rather as an evolutionary development.

More generally, these findings offer a novel and quite original understanding of wireless telegraph, one that could serve as the basis of a radical redefinition of the very notion of wireless and radio. Research into the terminology of the early twentieth century demonstrates that today's definitions of radio, telegraphy, broadcasting, telephone, and other media are somewhat rigid and inflexible. A more nuanced understanding of the various meanings and transitional identities of the media in question could help to identify their actual origins and clarify contentious issues in the discourse surrounding their invention. In particular, this research has demonstrated just how crucial it is to use the word 'radiotelegraphy', not simply because it features in the name itself of the IRU but also to address the controversy associated with this term in the histories of telecommunications and mass media. As Kate Lacey has emphasized, it is necessary for scholars to engage in dialogue 'beyond the confines of radio studies.'[5] Radiotelegraphy is itself a confine of precisely this kind: a medium that, for all its similarity to radio, predates it, and by the same token points the way to a wholly different narrative.

Research shows that radiotelegraphy represented an essentially different stage in media history and differed from wireless telegraphy and radio broadcasting. Defined in 1912 as a valuable instrument for ensuring safety at sea, radiotelegraphy was a common good to be shared by maritime nations. At its center, there were transnational collaborations, such as the time signals network, the meteorological network, and global-born corporations, which not only pursued imperial interests but also attempted to expand to different parts of the world by creating a coherent communication space. It was no longer simply a point-to-point network, in which wireless telegraphy was envisaged as a closed communication for private correspondence. It was also not yet a one-to-many system, which presupposed the asymmetrical spread of information. Instead, it was a

5 Kate Lacey, "Ten Years of Radio Studies: The Very Idea," *The Radio Journal* 6, no. 1 (2008): 21–32, https://doi.org/10.1386/rajo.6.1.21/4. 22.

transnational broadcasting system that allowed many stations to broadcast their messages at a specified time and be heard by many others across the globe.

I have attempted to highlight the differences between the historiography of wireless telegraphy and that of radio broadcasting (Table 3). The review of the literature revealed that the history of wireless telegraphy from 1890 to the 1910s is mainly based on the concept of point-to-point media, while radio broadcasting from the 1920s onwards is treated as a mass media tool for one-to-many communication. This research has by contrast demonstrated that the radiotelegraphy of the 1910s incorporated both point-to-point and one-to-many features, and in an altogether unique way. Stations across the globe were broadcasting time and weather reports at a particular hour to allow for flawless communication and also to avoid interfering with each other. Moreover, the stations that picked up this information could also communicate it, and therefore contribute to the spread of information by being active participants in this communication scheme rather than simply an audience.

Table 3: The main differences between the historiographies of wireless telegraphy, radiotelegraphy, and radio broadcasting.

	Radio	
Wireless telegraphy	**Radiotelegraphy**	**Radio broadcasting**
1890–1910s	1910s–1920s	The 1920s onwards
Point-to-point	Many-to-many	One-to-many
Long waves	Long and short waves	Long and short waves
Morse Code	Morse Code and Voice/music	Voice/music
Marconi Company, Société Française Radio- Electrique, Telefunken, etc.	Transnational collaborations: Debeg, time signals network, etc.	BBC, Radio Paris, Reichs-Rundfunk-Gesellschaft, etc.
Imperial	Transnational	National

The fact that radio messages were broadcast but yet also allowed two-way communication could be seen as another pattern of communication, which might be referred to as 'many-to-many.' Furthermore, transnational collaborations and projects played one of the most important roles in shaping the development of radiotelegraphy; it was not just a method serving national and imperial goals but was also used for the common good of all nations. In this way, radiotelegraphy could further be seen not just as a transitional stage between wireless telegraphy and radio broadcasting but also as a different medium with a set of unique characteristics.

Today we are on the threshold of the centenaries of the world's largest radio stations. The 2020s mark the centenary of radio in a number of different countries, with radio entering its "second century."[6] For instance, in 2022 the BBC, one of the first and best known public radio stations in Europe, celebrated its centenary. Nowadays, the BBC continues to be regarded as a benchmark for public service broadcasting worldwide, and it is not surprising that its birthday was seen by some as the birthday of radio as such.

This monograph both celebrates that birth of radio as a national medium in the 1920s and asserts that it has a longer and older history in the guise of radiotelegraphy. As not just a predecessor of radio broadcasting but also as a truly transnational technology designed to embrace the whole world, radiotelegraph networks built (and scheduled to be built) at the dawn of the twentieth century were, in fact, early predecessors of today's global media systems. Scholars claim that global communicative connectivity grew with the distribution of media products across different national borders and the emergence of the Internet,[7] however, the same claim could readily be made for the radiotelegraph in the 1910s. In its interconnectivity and transnational aspect, radiotelegraphy in the 1910s had much in common with the digital and global media systems of our day.

This is not the only instance of such continuities. Indeed, historians have drawn attention to numerous transnational phenomena arising in the analogue era. Meteorological networks,[8] the electrical telegraph,[9] and other communication technologies of early modernity could also be analyzed through the lens of interconnectivity and globalization. The electrical telegraph spread around the world as a web, reaching out to the most distant places in a manner astonishingly similar to the ubiquitous contemporary internet undersea cables, if comparing the maps of both networks.[10] Transnational communication and communicative connectivity in Europe were by no means new phenomena, and therefore it is therefore only logical that in these social and technological surroundings, radiote-

6 John Allen Hendricks, ed., *Radio's Second Century* (New Brunswick, Camden, and Newark, New Jersey, and London: Rutgers University Press, 2020).

7 e.g., Andreas Hepp, "Transculturality As a Perspective: Researching Media Cultures Comparatively," *Forum Qualitative Sozialforschung/Forum: Qualitative Social Research* 10, no. 1 (2009).

8 Edwards, "Meteorology as Infrastructural Globalism."

9 Simone M. Müller, *Wiring the World: The Social and Cultural Creation of Global Telegraph Networks* (New York, Chichester: Columbia University Press, 2016); Roland Wenzlhuemer, *Connecting the Nineteenth-Century World: The Telegraph and Globalization* (Cambridge: Cambridge University Press, 2012).

10 Maria Rikitianskaia and Paolo Bory, "Images of global communication: revealing maps of telecommunications infrastructure [Obrazy global'noj kommunikacii: raskryvaem karty telekommunikacionnyh infrastruktur]" (in Russian), *NZ* 130, no. 2) (2020), 40–54.

legraphy acquired the characteristics of global communication in the 1900s and 1910s. Radiotelegraphy was born as a global technology and was surrounded by various successful global projects in society. Unfortunately, its global reach was disrupted by the outbreak of the war.

In the long-term perspective, looking retrospectively at the history of radio, this means that transnationalism and global communication were at the origins of the technology and not the result of its development later. The projects of transnational radio broadcasting that sprang later, from the 1950s, have been researched as innovative in contrast to the national broadcasting practices of the 1920–1930s,[11] however, they could be reinterpreted not as innovative but as simply coming back to the roots of technology. That allows us to better understand media development within its longer history and to challenge what David Morley called 'the drastically foreshortened historical perspective' of media studies.[12]

Wireless then and now

Wireless surrounds us everywhere today. Wi-Fi and mobile networks allow continuous access to the Internet. Radio and television are omnipresent mass media delivered both through cables and by radio waves. Its communicative potentialities aside, wireless also helps in managing and coordinating various devices. Wireless keyboards, printers, mouses, earphones, stereo systems, and remote controls help to operate computers, smartphones, and television sets. Digital scales, TVs, vacuum cleaners, virtual assistants and indoor security cameras connect to our smartphones through wireless and by collecting data organize our lives better. It was once supposed that many household devices would never need a wireless connection, but total wireless connectivity appears to be integrated into more and more devices. Fridges nowadays are capable of telling us when we are running out of milk, connecting wirelessly to the internet and, recently, even ordering groceries online. Washing machines for their part are becoming 'smart' by connecting to our phones and allowing us to manage their rinsing cycles remotely. Even sneakers can connect to Wi-Fi, becoming self-lacing app-controlled shoes that need charging.

11 See, for example, Nelson Ribeiro and Stephanie Seul (eds.), *Revisiting transnational broadcasting: the BBC's foreign-language services during the Second World War* (London, New York: Routledge, 2017).

12 David Morley, *Media, Modernity and Technology: The Geography of the New* (London and New York: Routledge, 2007), 2.

Wireless today is so omnipresent that it often seems to be taken for granted. It is a curious fact that this contemporary wireless should have grown out of the early twentieth-century experiments with wireless telegraphy. There are, indeed, two histories of wireless: one of the early twentieth century, and another, more recent one. The best illustration of this is the frequency of the use of word 'wireless' in sources in English in Google's text corpora (see Figure 29). On the following chart, two peaks in the usage of word 'wireless' are clearly visible, one in the 1920s and another in the 2000s. Between these two peaks, the graph steadily plummets, probably because for long decades the term 'wireless' was buried under the popularity of radio broadcasting.[13] The steady rise in the use of the term 'wireless' commenced in the 1990s, with the development of wireless networks such as Wi-Fi and mobile networks. After having achieved a peak in the 2000s, the graph plunges downwards, probably on account of the development of multiple sophisticated wireless technologies, which are now mostly known by their own specific names.

Contemporary wireless practices surprisingly embody everything that enthusiasts of the 1910s and 1920s dreamed about: global connectivity, ease of connecting, the ubiquitous substitution of cable.[14] Today, our communication practices are already more wireless than ever, with the omnipresence of wireless technologies in becoming a precondition of most of our communication practices. 'Wirelessness,' observes Adrian Mackenzie, is a 'key form of contemporary existence'.[15] This 'wirelessness' refers not just to the ubiquity of the wireless around us, but also to a wider and deeper belief in media–technological progress. It is no longer enough for a contemporary fridge just to keep groceries, since nowadays it has to communicate progress wirelessly. Otherwise, it won't be modern enough.

Despite 100 years of difference in age, wireless of the twentieth and the twenty-first centuries have much in common. In Mackenzie's words, 'radio communication and information networks have an intimate historical association,'[16] and the contemporary wireless is not as innovative as we may suppose. There are a great many commonalities between these two wireless histories: the integration of

13 In general, throughout the twentieth century, there were many inventions that used the word "radio" in their name. Yet many inventions today either employ their own terminology, or use the word "wireless"; Maria Rikitianskaia and Gabriele Balbi, "Radio Studies beyond Broadcasting: Towards an Intermedia Radio History," *Radio Journal: International Studies in Broadcast & Audio Media* 18, no. 2 (2020).

14 'The day that is coming,' 1914.

15 Adrian Mackenzie, Wirelessness. Radical Empiricism in Network Cultures (Cambridge (MA), London: The MIT Press, 2010).

16 Adrian Mackenzie, *Wirelessness. Radical Empiricism in Network Cultures* (Cambridge (MA), London: The MIT Press, 2010), 4.

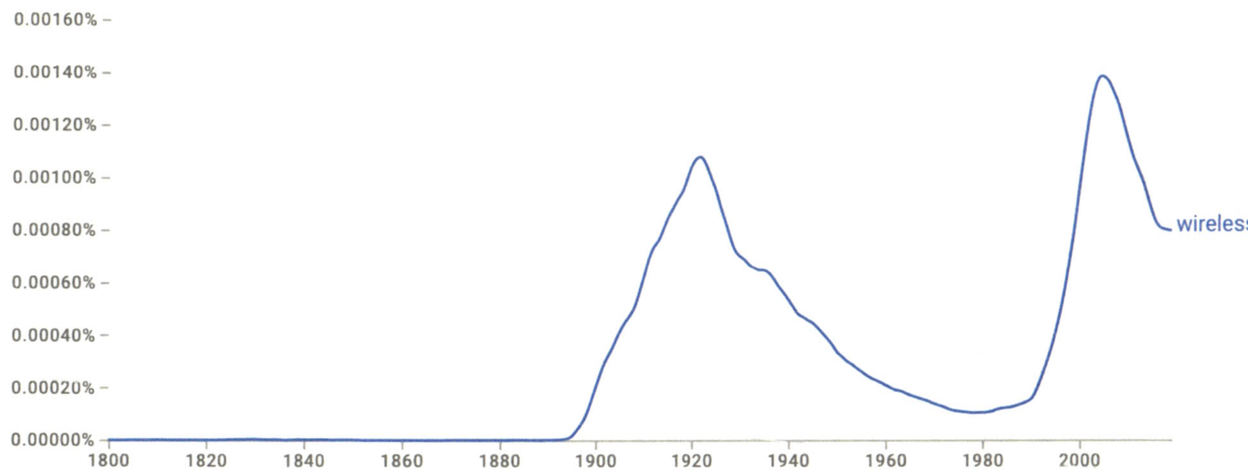

Figure 29: Frequency of the term ‘wireless’ over the years in sources printed between 1500 and 2019 in Google’s text corpora in English. Source: https://books.google.com/ngrams, accessed April 26, 2024.

wireless with cable infrastructure; the issue of security as it relates to radio waves and the imaginary of wireless as a secure channel from the Marconi presentations to Wi-Fi standardization; small scale wireless networks among amateurs then and now; and the persisting dilemmas regarding radio spectrum, spectrum holes, licenses, and other critical issues. With each new invention, national governments raised and addressed the question of the appropriate regulation of this new medium. Would it stay in private hands, be government-owned, or be under part public and part private ownership? To what extent should it be regulated and how? As these technologies emerged around the world, governments came up with a range of different approaches to their regulation, but in each case such choices influenced present and future telecommunications. The legacies of the wireless of the past are being handed on to contemporary practices.

Today, probably the most common association with the word 'wireless' is with Wi-Fi networks. Access to wireless Internet provided by Wi-Fi networks has had a huge impact on communication practices. From 2016 to 2018 alone, the number of Wi-Fi hotspots worldwide increased from 94 million to 169 million.[17] By 2023, 628 million public Wi-Fi hotspots were predicted to be available.[18] Even with mobile networks available, around 60% of the traffic of the mobile-connected devices came through fixed broadband with Wi-Fi access points in 2016, and Wi-Fi offload is likely to remain higher than 4G or 5G mobile networks due to mobile data caps.[19]

The advent of Wi-Fi was not an invention in its pure form, but rather a creation based on numerous other technologies. As Paul M. Torrens points out, 'Wi-Fi marries Internet-based networking and radio broadcasting.'[20] In fact, a Wi-Fi network is simply a sort of packet radio network, the idea being to send packets of data via radio communication. Creating packet radio networks has been a common practice for radio amateurs since at least the 1910s.[21] Wi-Fi was only a logical continuation of something else, involving the application of an already existing practice to new information and communication systems.

Wi-Fi's creation was driven by the release of the industrial, scientific, and medical (ISM) radio bands for unlicensed use by the U.S. Federal Communications Com-

17 Cisco, "Cisco Visual Networking Index (VNI) [White Paper]" (2017), 20; Cisco, "Cisco Annual Internet Report (2018–2023). White Paper" (2020), 13.

18 Cisco, "Cisco Annual Internet Report (2018–2023). White Paper," 13.

19 Cisco, "Cisco Visual Networking Index (VNI) [White Paper]," 17–19.

20 Paul M Torrens, "Wi-Fi Geographies," *Annals of the Association of American Geographers* 98 (2008): 59.

21 Streeter, *Selling the Air: A Critique of the Policy of Commercial Broadcasting in the United States*, 75.

mission in 1985. In the decade after the ISM release, several wireless networks were introduced to the market, such as Token-Ring and WaveLAN. In 1990, The Institute of Electrical and Electronics Engineers formed the 802.11 Working Group for Wireless LANs to standardize wireless networks, and in 1997 the committee issued the original open IEEE 802.11 standard.[22] By the late 1990s, many technology companies had accommodated Wi-Fi connections in their devices; e.g., Apple incorporated wireless cards into every Apple product, including PowerBooks, eMacs, iMacs, etc. The technology grew quickly in the early 2000s and proliferated in many everyday practices. As a smart marketing move, the name Wi-Fi was coined to resemble the acronym Hi-Fi (high fidelity), a generic term used to indicate high-quality audio technologies. Contrary to common belief, Wi-Fi is not in fact an acronym for wireless fidelity, yet it worked very well. In 2003, the official magazine of the International Telecommunication Union highlighted its rapid growth with an article entitled, 'Wi-Fi takes the (communication) sector by storm.'[23]

The Wi-Fi standard was created to allow the communication market to use the ISM bands to full advantage within their specific limitations, paired with the use of the previously secret military technology of spread spectrum. The most important purpose was to offer ease of connection and portability of devices, while still maintaining the level of speed of wired networks. Interestingly, the original purpose of Wi-Fi was not to connect computers, as they were massive, immovable, and could already be connected perfectly using wired LANs. The first customers for Wi-Fi standard came from the retail industry, as wireless networks were designed to connect cash machines.[24] The latter, unlike mainframes, had to be more portable to respond to the needs of merchandising. Department stores were constantly rearranging their collections, and therefore moving their transaction terminals. Rewiring them was highly inconvenient, and connection via radio waves was thus essential. That is the reason why the most important actor in Wi-Fi development was the National Cash Register (NCR) Corporation, which was known for manufacturing and selling the first mechanical cash register in the late nineteenth century.

This history of Wi-Fi encapsulates an important issue for wireless: the need to always be interconnected to the wired networks. On the one hand, the development of wireless naturally entails a degree of antagonism towards wired communication. Its creation always comes a bit later than the corresponding wired connections, and

22 Shane Greenstein, "Concentration in Internet Access and Entrepreneurial Truncation of Innovation," *Capitalism and Society* 7, no. 1 (2012): 19, https://doi.org/10.1515/1932-0213.1092.

23 ITU News, "Wi-Fi Takes the Sector by Storm," *ITU News* 8, no. 8 (2003): 29–31.

24 Maria Rikitianskaia, ""The Real Ethernet": The Transnational History of Global Wi-Fi Connectivity," *New Media & Society* (2022): 1–20, https://doi.org/10.1177/14614448221103533.

it thus has to compete in terms of speed or ease of connection, and also to address very specific customers. Just as wireless telegraphy was first appreciated at sea, because it was helpful for navigating ships in motion, so too was Wi-Fi created for movable cash registers.[25] On the other hand, wireless relies on wired networks and is designed to be integrated into a large communication infrastructure. The pioneers in the field of wireless innovation are often already established actors in the wired market. At the outset wireless telegraphy was integrated into postal and telegraph offices, being simply another option on offer. Wi-Fi networks, in their turn, spread thanks to the efforts of telecom industry players, such as British Telecom, which offers the full package of wired Internet access plus a modem to offices and private homes alike. In a way, this competition with wired analogues and the general tendency towards integration also forces those engaged in wireless innovation to mimic the features of older technologies, as happens to many old and new media.[26]

It is impossible not to mention the phenomenally rapid development of the mobile phone and its proliferation within our communication practices today, with it making up a huge part of our wireless connectedness and contributing greatly to the notion of the global communication space. In fact, the development of the mobile phone corresponds well with the early twentieth-century attempts to improve the experience of using wireless telegraphy. As discussed throughout this book, wireless telegraphy was imagined mostly as a form of point-to-point communication rather than a one-to-many broadcasting medium. Marconi himself imagined the future of wireless telegraphy in terms of a call from France to Italy, and as such resembling the use of the mobile phone today:

> Pocket wireless telephones by means of which a passenger flying in an aeroplane over France or Italy might "ring up" a friend walking about the streets of London with a receiver in his pocket cannot be said to have been as yet practically realized but there is nothing inconceivable or even impracticable about such an achievement and the progress of wireless telephony seems to be pointing in that direction.[27]

The proliferation of mobile phones today echoes that future envisioned by Marconi for his invention. Their history could be said to start with the invention of the first walkie-talkies in 1937, as two-directional portable transmitters and receivers. The next step in the development of mobile telephony was the idea of the network set by multiple antennas, called cell towers, that could connect people by

25 In general, the retail industry has been one of the first adopters of wireless innovations. Thus, department store stations featured very prominently in the radio broadcasting boom of the 1920s.
26 Balbi, "Old and New Media. Theorizing Their Relationships in Media Historiography."
27 Cited by Raboy, *Marconi: The Man Who Networked the World*, 382.

means of radio waves. The forerunners were national mobile phone networks, coming from Northern Europe and Switzerland. Originally being quite heavy and bulky, the mobile phones were themselves installed in cars. Even though switching from one tower to another was not easy, the networks provided an exceptional connection while on the move. Companies even provided a specific map to let their subscribers know how to call someone depending on which part of their country they were located in. However, the very first cell phone network dates back to 1979, having been launched in Tokyo, Japan, for public use. Despite having developed as national projects, the mobile networks soon called for the creation of a unified communication space, which did not happen until the 1990s. In Europe, the common standard for mobile telephony was adopted in 1992 by eight European countries, thus setting an example for the whole world.

Another continuity in wireless communication between the early twentieth century and the contemporary period is the use of the binary code. Today, we are quite accustomed to a discourse predicated on digitalization, which is based on coding information into digits. We tend to forget that those digits are themselves binary – combinations of zeros and ones, which so readily permit the decoding of all information. History offers many examples of binary codes which are just as powerful. Morse code, used both for wired and wireless telegraphy in the nineteenth and twentieth century, is probably the most influential binary code; the combinations of dots and dashes can encode any information whatsoever, just as zeros and ones do today. Wireless telegraphy transmitted texts, images, sounds, music, games, even x-rays. The accuracy of transmission did apparently leave much to be desired, and yet those transmissions did take place and also improved with time. The idea itself – to decode information, transmit through the radio waves, and encode on the machine – was absolutely the same 100 years ago as it is today.

Another common issue for wireless innovations is their relation to the radio spectrum. Just as 100 years ago, today the international organization responsible for the regulation of the radio spectrum is still the ITU. However, at the beginning of the twentieth century, the division of the radio spectrum was far from being as elaborate as it is now. Starting from the first IRU conferences in the 1900s, countries assigned individual bands to specific purposes. Even a quick glance at the two images of the radio spectrum today and 100 years ago reveals how elaborate and intricate the division of radio has become. All the pieces of the different radio services form part of a bizarrely whimsical mosaic nowadays. In fact, the biggest problem for contemporary wireless innovation is the lack of an available radio spectrum, with some contemporary wireless practices even more peculiar because they go beyond that radio spectrum division. The Wi-Fi networks, for instance, developed outside the scope of the licensed radio spectrum, and they are still growing today in the 'grey area' of international regulations, causing a kind of 'radio revolution.'[28] That is why the 5G networks, for instance, use a very complex combination of different parts of the radio spectrum; in addition to the bands previously allocated to mobile communication, 5G also appropriates some unlicensed parts. The struggle for the radio spectrum becomes fiercer with the release of every new wireless product.

Today, the development of that part of the spectrum is now at the center of multiple controversies. In the era of the developing of Wi-Fi, the main advocate of spread-spectrum technology, Michael Marcus, then working for the US Federal Communications Commission, observed little resistance to using the higher end of the spectrum for wireless networking. 'Microwave ovens,' Marcus wryly remarked, 'don't protest.'[29] It is interesting to note how that relationship with microwave ovens has changed over time, as the controversies over 5G development indicate. The 5G networks are designed to use a part of the radio spectrum originally allocated for non-communication purposes; thus, their position is not so very different from that of the globally approved and uncontroversial Wi-Fi standard.[30] However, the fact that 5G uses the part of radio spectrum adjacent to the microwave one is often a crucial point for critics who have health concerns and fears about radiation.

The fears provoked by new technology have much in common with the anxieties aroused by every innovation, and here we can plausibly compare the reception granted to wireless telegraphy with the response given to contemporary wireless products. Indeed, wireless telegraphy was the object of fear for many decades. Communication without any wires and any visible connection was associated with spiritualism, mind-reading, and contact with the afterlife;[31] numerous people wrote to Marconi accusing him personally of reading their minds. At a good historical distance, these accusations seem almost comical, especially given how the word 'wireless' is used so widely when marketing new products, the im-

---

28 Kevin Werbach, *Radio Revolution: The Coming Age of Unlicensed Wireless* (Washington: New America Foundation Public Knowledge, 2003).

29 As cited in Wolter Lemstra, Vic Hayes, and John Groenewegen, eds., *The Innovation Journey of Wi-Fi: The Road to Global Success* (Cambridge: Cambridge University Press, 2011), 372, https://doi.org/10.1017/CBO9780511666995.

30 Wolter Lemstra, "Leadership with 5 G in Europe: Two Contrasting Images of the Future, with Policy and Regulatory Implications," *Telecommunications Policy* 42, no. 8 (2018): 587–611, https://doi.org/10.1016/j.telpol.2018.02.003.

31 Natale, *Supernatural Entertainments: Victorian Spiritualism and the Rise of Modern Media Culture*; David Hendy, "The Dreadful World of Edwardian Wireless," in *Moral Panics, Social Fears, and the Media: Historical Perspectives. Routledge Research in Cultural and Media Studies* (London and New York: Routledge, 2013).

plication being that they are dazzlingly progressive and modern. Yet closer to our own time, there are highly convincing parallels with contemporary fears of wirelessness. The frequent discussions of health concerns because of 5G could be seen as moral panics similar to those erupting in the past. Moreover, the current circumstance of Covid-19 has led to accusations being levelled at mobile providers, as if 5G antennas were transmitting the virus or reducing the human defences to it.[32]

The issue of radio spectrum brings to the fore yet another continuity in wireless evolution: the use of spectrum holes. For instance, this is one of the innovative features of the 5G networks, stemming, once again, from the lack of available spectra. A spectrum hole is a licensed part of the spectrum which is not being used under certain circumstances. Thus, the goal of 5G is to utilize those available spaces whenever they are vacant. Even though at that point the radio spectrum was not as busy as it is now, wireless telegraphy also created spectrum holes for a very specific message: the emergency SOS call. The 1912 idea to drop all communication and to listen to the ether for a couple of minutes not only brought up a new idea for the organization of radio communication, but also created a precedent for the future use of radio. Those holes and gaps in radio spectrum matter even more today.

The abovementioned release of those ISM bands in 1985 drove the development of many low-power wireless communications. For instance, one of the crucial wireless innovations is the Bluetooth standard, which allows point-to-point connection within a small radius. Earphones, mouses, and keyboards connect via Bluetooth to our devices, with this wireless technology standard released in 1999 to boost the connection among various wireless devices (computers, phone, radio, etc.) that were commonly characterized by incompatible communication protocols. Recently, another new and important application of Bluetooth was revealed, with Apple and Google jointly releasing a contact-tracing function via Bluetooth during the Covid-19 pandemic.[33] The idea was to allow smartphones to log all the available information about any devices in the immediate vicinity, so as to record all contacts that may have occurred and be able to trace the spread of the virus; if a user marked themselves as infectious, their past contacts were notified through the app.[34] The function was not only incorporated into the official national apps to combat the virus, but also became integrated into new iOS and Android operating

32 Jack Goodman and Flora Carmichael, "Coronavirus: 5 G and Microchip Conspiracies around the World," BBC News (2020), https://www.bbc.com/news/53191523.

33 See, for instance, the official British app NHS Covid.

34 Apple Inc., "Privacy-Preserving Contact Tracing" (2020), https://covid19.apple.com/contacttracing.

systems. Constant wireless connectivity in this case became a tool for population surveillance – which in this circumstance should allegedly serve the public good.

This new application of wireless connectivity also reveals another parallel with the discussions about wireless at the beginning of the twentieth century. The contact-tracing function of our smartphones renders it possible for our connections to be tracked, whether we wish it or not, and without us noticing. There is no need to log in specifically, or to establish specific connections, since the devices will automatically track everything that is around.[35] It is technically possible because wireless waves know no barriers; once the antenna is turned on, everything capable of receiving the signal in its radiation radius will be able to catch it. In a certain sense, wireless waves are unstoppable, with this constant penetration by wireless already regarded as problematic at the beginning of the twentieth century. The advent of wireless telegraphy therefore required additional regulation of the airspace, as the waves could hardly be restricted; some suggestions to prevent waves from entering national territories included such far-fetched ideas as building a 'high metal wall' or erecting a 'high net of iron wire.'[36] The politicians were trying to decide whether an airspace should be considered a national or an international space, but in any event the advent of wireless led to there being a vertical division of the air.[37]

In addition, throughout the entire course of wireless history, there has been a constant confrontation between top-down political and economic decisions and bottom-up amateurish initiatives. As this monograph has demonstrated, radio amateurs sprang from the scattered community of technology enthusiasts experimenting with wireless telegraphy in the 1900s and 1910s. Individualistic though such experiments were, they led to the proliferation of wireless and radio in ordinary households, and to these new devices becoming an integral part of everyday life. From picking up news and weather information to broadcasting music and concerts, the content produced by amateurs occupied an important part of the radio. They demonstrated that wireless could be not only a strategic tool deployed by postal or military authorities, but a fun medium for the whole family and neighbourhood, with this appropriation of technology by amateurs a common feature in the history of many different technologies. Similarly, the proliferation of personal computers famously owes much to experiments conducted by enthusiasts in their

35 It is in fact possible to turn off this service, however, by default it is constantly on.

36 Torquato Carlo Giannini, *La Radiotelegrafia Nell'economia e Nella Legislazione* (Roma: Agenzia Radiotelegrafica Italiana, 1920), 41.

37 Rikitianskaia, Balbi, and Lobinger, "The Mediatization of the Air. Wireless Telegraphy and the Origins of a Transnational Space of Communication, 1900–1910s."

garages or workshops. Steve Jobs, together with Steve Wozniak, started Apple Inc., now a multinational company, in the garage of a modest house in California. Similarly, just 10 miles from the Jobs garage, the Hewlett-Packard Company sprang from another tech start-up garage. These astonishing successes and big names aside, the activities of amateurs and enthusiasts are also important for wireless development on a small scale. For instance, one of the more marginal activities of radio amateurs is connected to the creation of wireless networks. Trying to retain the vision and organizing strategies characteristic of low power FM radio initiatives, activists today create community Wi-Fi networks, thereby extending their sense of community media into internet-based technologies.[38] Technical enthusiasts set up networks and commonly reconfigure their wireless access points, thereby modifying default settings, which can be seen as a powerful form of regulation.[39]

Finally, it important to remember that all types of wireless media no matter how powerful and modern they may be, still require a certain materiality of infrastructure. And this infrastructure has typically become an enduring legacy. For instance, 5G antennas are often mounted where mobile antennas of previous generations had been. Using a site where existing communications are located is simply convenient for telecommunication engineers; for this reason, 5G antennas are often situated where 4G, GSM, and satellite dishes are already installed. Not surprisingly, when attempting to destroy 5G phone masts in accordance with the view that they form part of a conspiracy to spread coronavirus, vandals have actually damaged 3G or 4G equipment.[40]

No matter how ephemeral, invisible, and intangible radio waves are, they always leave a significant material trail or trace. The invisible and hidden nature of

38 Christina Dunbar-Hester, '"Free the Spectrum!" Activist Encounters with Old and New Media Technology', *New Media and Society* 11, no. 1–2 (2009): 221–40, https://doi.org/10.1177/1461444808100160; Stefano Crabu and Paolo Magaudda, "Bottom-up Infrastructures: Aligning Politics and Technology in Building a Wireless Community Network", *Computer Supported Cooperative Work: CSCW: An International Journal* (2017), https://doi.org/10.1007/s10606-017-9301-1; Alison Powell, "WiFi Publics: Producing Community and Technology", *Information Communication and Society* (2008), https://doi.org/10.1080/13691180802258746.

39 Laura Forlano, "WiFi Geographies: When Code Meets Place," *Information Society: An International Journal* 25, no. 5 (2009): 344–52, https://doi.org/10.1080/01972240903213076; Rajiv C. Shah and Christian Sandvig, "Software Defaults as de Facto Regulation: The Case of the Wireless Internet," *Information Communication and Society* 11, no. 1 (2008): 25–46, https://doi.org/10.1080/13691180701858836.

40 Jim Waterson and Alex Hern, "At Least 20 UK Phone Masts Vandalised over False 5 G Coronavirus Claims," *The Guardian*, April 6, 2020, https://www.theguardian.com/technology/2020/apr/06/at-least-20-uk-phone-masts-vandalised-over-false-5g-coronavirus-claims.

radio waves could be better understood in the context of a war that necessitated the disruption of infrastructures; radiotelegraphy as an 'invisible weapon'[41] was not just invisible but also bulky and tangible. This monograph has drawn attention to the devastating consequences of destroying high antennas and stations at the outbreak of war, and also to the clothes line as a possible way of masking an antenna, which became both a sign for radio communication and the object of fear. In other words, this perspective on the medium during one of the tragic periods of its destruction serves to reveal its usually obscured material infrastructure. During World War I, radio communication antennas were blown up or dismantled, and their remnants have become a kind of legacy. By the same token, today's antennas will also one day become relics of an outmoded past. Today, scholars acknowledge that the materiality of infrastructures does indeed play an essential role in global communications and some even go so far as to claim that, by contrast with the wireless past, the future will be surprisingly 'wired.'[42] Behind wireless communication there is a very advanced material infrastructure, which is being developed in steps and is now contributing to the increase of digital waste in the world.

Radio waves carry information. The air was already full of information even a hundred years ago, but today it is densely packed with it. In line with recent mediatization theories, this process of appropriating the air as a communication space could be called the mediatization of the air.[43] Thanks to radio waves, the air today is a medium permitting our inclusion within the global communication space.

At the beginning of the twentieth century, people could only dream of the successful realization and spread of the most impressive wonders of wireless. One of the lists of possible applications of wireless, published in 1914, was called 'The Day That Is Coming'. It included radiolocating, listening to concerts, sending pictures from Berlin to Paris, photographing lions, regulating lighting, and exploding mines by means of wireless.[44] Today, most of these anticipations of the miraculous powers of wireless have been fully realized and, thus, the greatest hopes of the innovators of the early twentieth century have been fulfilled. The

41 Headrick, *The Invisible Weapon: Telecommunications and International Politics, 1851–1945.*

42 Ghislain Thibault, "Wireless Pasts and Wired Futures," in *Theories of the Mobile Internet: Materialities and Imaginaries*, ed. Andrew Herman, Jan Hadlaw, and Tom Swiss (New York, Oxon: Routledge, 2015), 87–104.

43 Rikitianskaia, Balbi, and Lobinger, "The Mediatization of the Air. Wireless Telegraphy and the Origins of a Transnational Space of Communication, 1900–1910s."

44 "The Day That Is Coming," *My Children's Magazine*, October 1914, 135–43.

potent and highly influential vision of global connectivity by wireless remained for decades in the imaginations of those inventors and engineers who worked out different ways to connect the world. Our communication practices today are already more wireless than ever. Recasting the title of the 1914 article, the day has come.

References

"A Brief Account of What Happened at Buffalo . . ." *QST* 7 (1917): 26.

"A Curious Misunderstanding." *The Wireless World* (April 1916): 32.

"A Lady Telegraphist." *The Wireless World* (October 1917): 475–77.

"A Novel Fact That We Announced . . ." *The Wireless World* (March 1914): 749.

Adair, John. "The Marconi Contract." *The Commentator*, January 22, 1913.

Aitken, Hugh G.J. *Syntony and Spark. The Origins of Radio*. 1985.

Ala'i, Padideh, and Robert G Vaughn, eds. *Research Handbook on Transparency*. Cheltenham, Northampton: Edward Elgar, 2014.

Alekseev, Timofej Vladimirovich. *Communication Industry of Petersburg-Leningrad for the Army and Navy in the Era of Commotions and Modernization, 1900–1945 [Industrija Sredstv Svjazi Peterburga-Leningrada Dlja Armii i Flota v Jepohu Potrjasenij i Modernizacii, 1900–1945 Gody]*. Saint-Petersburg: Peter the Great St. Petersburg Polytechnic University, 2010.

Alfani, Guido Padre. "Estratto: L'Osservatorio Ximeniano e Il Suo Materiale Scientifico: V (La Stazione Radiotelegrafica). Firenze : Tip. S. Giuseppe, 1912." *Rivista Di Fisica, Matematica e Scienze Naturali – Pisa* 13 (1912): 148.

Allinson, Caroline. "The Process of Audit and Control – A Comparison of Manual and Electronic Information Systems." *Policing: An International Journal of Police Strategies & Management* 27, no. 2 (2004): 183–205. https://doi.org/10.1108/13639510410536814.

Altman, Rick. *Silent Film Sound*. New York: Columbia University Press, 2005.

"Amateur Work during the War." *The Wireless World* (August 1915): 322–23.

"An Aerogram for Vienna." *The Wireless World* (September 1918): 340.

"An Interesting Lecture." *The Wireless World* 12 (1914): 577.

"An International Amateur Wireless Congress for 1925." *The Wireless World and Radio Review* (April 1924): 137.

Anderson, Benedict. *Imagined Communities: Reflections on the Origin and Spread of Nationalism*. London and New York: Verso, 1983.

Andogsky, Alexander Ivanovich. *Meeting Engagement. Strategic and Tactical Investigation of Ways and Methods of Conducting Meetings of a New Battle in the Modern Era on the Basis of Military-Historical Examples [Vstrechnyj Boj. Strategichesko-Takticheskoe Izsledovanie Sposobov i Priemov*. Petrograd: General Staff Academy, 1918.

Apple Inc. "Privacy-Preserving Contact Tracing." 2020. https://covid19.apple.com/contacttracing.

"Application de La Radiotélégraphie. Avertissements Météorologiques." *Journal Télégraphique* 10 (1908): 324.

Arceneaux, Noah. "The Wireless Press and the Great War: An Intersection of Print and Electronic Media, 1914–1921." *Journal of Radio & Audio Media* 26, no. 2 (2019): 318–35. https://doi.org/10.1080/19376529.2018.1497035.

"Argentine." *Journal Télégraphique* 1 (1918): 16.

"Article 45." In *International Radiotelegraph Convention*, 206. London: His Majesty's Stationery Office, 1912.

Badenoch, Alexander. "Myths of the European Network: Construction of Cohesion in Infrastructure Maps." In *Materializing Europe: Transnational Infrastructures and the Project of Europe*, 47–77. Basingstoke, Hampshire and New York: Palgrave Macmillan, 2010. https://doi.org/10.1057/9780230292314_4.

Open Access. © 2024 the author(s), published by De Gruyter. BY-NC-ND This work is licensed under the Creative Commons Attribution-NonCommercial-NoDerivatives 4.0 International License.

https://doi.org/10.1515/9783111202327-007

Badenoch, Alexander, and Andreas Fickers. "Introduction: Europe Materializing? Toward a Transnational History of European Infrastructures." In *Materializing Europe: Transnational Infrastructures and the Project of Europe*, 1–26. Basingstoke, Hampshire and New York: Palgrave Macmillan, 2010.

Badenoch, Alexander, Andreas Fickers, and Christian Henrich-Franke. "Airy Curtains in the European Ether: Introduction." In *Airy Curtains in the European Ether. Broadcasting and the Cold War*, edited by Alexander Badenoch, Andreas Fickers, and Christian Henrich-Franke. Baden-Baden: Nomos, 2013.

Baker, William John. *A History of the Marconi Company*. London, 1970.

Balbi, Gabriele. *La Radio Prima Della Radio. L'araldo Telefonico e l'invenzione Del Broadcasting in Italia*. Rome: Bulzoni, 2010.

Balbi, Gabriele. "Marconi's Diktats. How Italian International Wireless Policy Was Shaped by a Private Company, 1903–1911." In *3rd Region-8 IEEE HISTory of Electro – Technology CONference: The Origins of Electrotechnologies, HISTELCON 2012 – Conference Proceedings*. 2012. https://doi.org/10.1109/HISTELCON.2012.6487559.

Balbi, Gabriele. "Old and New Media. Theorizing Their Relationships in Media Historiography." In *Theorien Des Medienwandels*, edited by Susanne Kinnebrock, Christian Schwarzenegger, and Thomas Birkner Hrsg, 231–49. Köln: Herbert von Halem Verlag, 2015.

Balbi, Gabriele. "Telecommunications." In *Handbook of Communication History*, edited by Peter Simonson, Janice Peck, Robert T. Craig, and John Jackson, 209–22. New York and London: Routledge, 2013.

Balbi, Gabriele. "Wireless's "Critical Flaw": The Marconi Company, Corporation Mentalities, and the Broadcasting Option." *Journalism & Mass Communication Quarterly* 1, no. 22 (2017): 1–22.

Balbi, Gabriele, Simone Fari, Giuseppe Richeri, and Spartaco Calvo. *Network Neutrality: Switzerland's Role in the Genesis of the Telegraph Union, 1855–1875*. Bern: Peter Lang, 2014.

Balbi, Gabriele, and Juraj Kittler. "One-to-One and One-to-Many Dichotomy: Grand Theories, Periodization, and Historical Narratives in Communication Studies." *International Journal of Communication* 10 (2016): 1971–90.

Balbi, Gabriele, and Simone Natale. "The Double Birth of Wireless: Italian Radio Amateurs and the Interpretative Flexibility of New Media." *Journal of Radio and Audio Media* 22, no. 1 (2015): 26–41. https://doi.org/10.1080/19376529.2015.1015860.

Barnard, C.C. "A Morse Practice Set." *The Wireless World* (November 1914): 528.

Bartlett, Richard A. *The World of Ham Radio, 1901–1950: A Social History*. Jefferson, London: McFarland, 2007.

Bayly, C.A., Sven Beckert, Matthew Connelly, Isabel Hofmeyr, Wendy Kozol, and Patricia Seed. "AHR Conversation: On Transnational History." *The American Historical Review* 111, no. 5 (December 2006): 1441–64.

Beauchamp, Ken. *History of Telegraphy*. London: The Institution of Engineering and Technology, 2008.

Belar, Albin. *Ein Erinnerungsblatt, Gewidmet Der 'Società Sismologica Italiana' Zu Ihrem 10 Jährigen Gründungsfeste*. Laibach: Kleinmayr & Fed. Bamberg, 1905.

Bijker, Wiebe E. "How Is Technology Made?-That Is the Question!" *Cambridge Journal of Economics* 34, no. 1 (2009): 63–76. https://doi.org/10.1093/cje/bep068.

Birchall, Clare. "Introduction to "Secrecy and Transparency": The Politics of Opacity and Openness." *Theory, Culture & Society* 28, no. 8 (2011): 7–25. https://doi.org/10.1177/0263276411427744.

Björnsson, Helgi. "Radio-Echo Sounding Maps of Storglaciären, Isfallsglaciären and Rabots Glaciär, Northern Sweden." *Geografiska Annaler. Series A, Physical Geography* 63, no. 3/4 (1981): 225.

Boczkowski, Pablo, and Leah A. Lievrouw. "Bridging STS and Communication Studies: Scholarship on Media and Information Technologies." In *The Handbook of Science and Technology Studies*, edited by E.J. Hackett, O. Amsterdamska, M. Lynch, and J. Wakcman, 949–77. Cambridge (MA), London: The MIT Press, 2008.

Bourdon, Jérôme. "Live Television Is Still Alive: On Television as an Unfulfilled Promise." *Media, Culture and Society* 22, no. 5 (2000): 531–56. https://doi.org/10.1177/016344300022005001.

Bown, Alfie. *Enjoying It: Candy Crush and Capitalism*. John Hunt Publishing, 2015.

Brecht, Bertolt. "Radio as a Means of Communication. A Talk on the Function of Radio." *Screen* 20, no. 3–4 (1979): 24–28.

Brown, Barry, and Oskar Juhlin. *Enjoying Machines*. Cambridge (MA), London: The MIT Press, 2015.

Brüggemann, Michael, and Hagen Schulz-Forberg. "Becoming Pan-European? Transnational Media and the European Public Sphere." *The International Communication Gazette* 71, no. 8 (2009): 693–712.

Bruton, Elizabeth Mary. "Beyond Marconi: The Roles of the Admiralty, the Post Office, and the Institution of Electrical Engineers in the Invention and Development of Wireless Communication up to 1908 (PhD Thesis)." The University of Leeds, 2012.

Bruton, Elizabeth Mary. ""Uncertain at Present for Women, but May Increase": Opportunities for Women in Wireless Telegraphy during the First World War." *Information & Culture* 55, no. 1 (2020): 51–74. https://doi.org/10.7560/IC55104.

Bruton, Elizabeth Mary, and Graeme Gooday. "Listening in Combat – Surveillance Technologies beyond the Visual in the First World War." *History and Technology* 32, no. 3 (2 July 2016): 213–26. https://doi.org/10.1080/07341512.2016.1231782.

BT Archives. "POST 30/2424. Experiments and Licences for Galletti Wireless Telegraph System, Part 1." London: BT Archives, 1915.

BT Archives. "POST 30/2425A. Experiments and Licences for Galletti Wireless Telegraph System, Part 2." London, 1914.

BT Archives. "POST 30/2576B. Transmission by Wireless Telegraphy of Time Signals and Weather Reports (Change to 1913–1914)." London, 1913.

BT Archives. "POST 30/2589B. Employment of Women as Wireless Telegraph Operators. 1913–1920." 1915.

BT Archives. "POST 30/2795C. Railophone System for Wireless Communication with Moving Trains (1910–1917)." London: BT Archives, 1917.

BT Archives. "POST 30/2921B International Telegraph Union Informal Conferences of Technical Experts." 1914.

BT Archives. "POST 30/3089A International Telegraph Conference, Paris in 1915, Part 2." 1915.

BT Archives. "Records of the Wireless Telegraphy Establishment; 1910–2001. Administrative History." 1910.

Bull, Stephen, and Adam Hook. *World War I Trench Warfare (1): 1914–16*. Oxford: Osprey Publishing, 2002.

Bureau des longitudes. *Conférence Internationale de l'Heure*. Paris: Gauthier-Villars, 1912.

Bureau des longitudes. "Electro-Optique (Suite)." In *Annuaire Pour l'an 1916 Publié Par Le Bureau Des Longitudes*, 454. Paris: Gauthier-Villars et Cie, 1916.

Bureau des longitudes. *Wireless Time Signals. Radio-Telegraphic Time and Weather Signals Transmitted from the Eiffel Tower, and Their Reception [Authorised Translation]*. London: E. & F. N. Spon, Ltd, 1915.

Burns, Russell W. *Communications: An International History of the Formative Years*. IET, 2004.

Caroll, Earl. *Send Me a Kiss by Wireless [Song]*. New York: Jos. W. Stern & Co, 1911.

Cartwright, William. "Rethinking the Definition of the Word "Map": An Evaluation of Beck's Representation of the London Underground through a Qualitative Expert Survey." *International Journal of Digital Earth* 8, no. 7 (2015): 522–37. https://doi.org/10.1080/17538947.2014.923942.

Castells, Manuel. *The Information Age: Economy, Society, and Culture. Volume I. The Rise of the Network Society*. Oxford: Blackwell, 1996.

Cejtlin, V.M. *Intelligence Service Staff [Razvedyvatel'naja Sluzhba Shtabov]*. Cmolensk: Upravlenie voenno-uchebnyh zavedenij zapadnogo fronta, 1921.

Christensen, Clayton M. *The Innovator's Dilemma: When New Technologies Cause Great Firms to Fail. The Innovator's Dilemma*. Boston, MA: Harvard Business Review Press, 1997.

Christensen, Miyase. "Trans National Media Flows : Some Key Questions and Debates." *International Journal of Communication* 7 (2013): 2400–18.

"Chronicle [Hronica]." *Herald of Military Radio Telegraphy and Electrical Engineering [Vestnik Voennoj Radiotelegrafii i Jelektrotehniki]* 3 (1917): 132–33.

Cisco. "Cisco Annual Internet Report (2018–2023). White Paper." 2020.

Cisco. "Cisco Visual Networking Index (VNI) [White Paper]." 2017.

Clarricoats, J. *World at Their Fingertips. The Story of Amateur Radio in the United Kingdom and a History of the Radio Society of Great Britain*. London: Radio Society of Great Britain, 1967.

"Code Radiotélégraphique International." *Journal Télégraphique* 11 (1919): 183.

Collins, Francis A. "Delivering a Wireless Message." In *The Wireless Man: His Work and Adventures on Land and Sea*. New York: The Century Co, 1912.

"Conférence Internationale Sur Les Communications Électriques." *Journal Télégraphique* 12 (1922): 259.

Cox, Jim. "Vincent Lopenz. The Piano Kid." In *Musicmakers of Network Radio: 24 Entertainers, 1926–1962* , 210–24. Jefferson: McFarland, 2012.

Crabu, Stefano, and Paolo Magaudda. "Bottom-up Infrastructures: Aligning Politics and Technology in Building a Wireless Community Network." *Computer Supported Cooperative Work: CSCW: An International Journal* (2017). https://doi.org/10.1007/s10606-017-9301-1.

Crisell, Andrew. *An Introductory History of British Broadcasting*. London and New York: Routledge, 1997.

Crisell, Andrew. "General Introduction." In *More than a Music Box: Radio Cultures and Communities in a Multi-Media World*. New York and Oxford: Berghahn Books, 1996.

Crocker, Irving, and George L. Cobb. *Send Me a Line When I'm across the Ocean [Song]*. New Yor: Walter Jacobs, 1917.

DAHR. "OKeh Matrix S-70586. Kiss Me by Wireless / Hotel Pennsylvania Orchestra; Vincent Lopez." Discography of American Historical Recordings, 1922. https://adp.library.ucsb.edu/index.php/matrix/detail/2000202719/S-70586-Kiss_me_by_wireless.

Daily Herald. "Women for Wireless." April 8, 1912. https://doi.org/10.1177/095574900001200206.

Daily Sketch. "Wireless Station for 30s." February 6, 1913.

Danic, Th. Le. *La Conférence Télégraphique Internationale de Paris (1925)*. Paris, 1926.

Das, Santanu. "Introduction." In *Race, Empire and First World War Writing*, edited by Santanu Das. Cambridge, New York: Cambridge University Press, 2011.

Department of Commerce Washington. "D. 40. N. 25." In *Correspondance: Radiotélégraphie*. ITU Library & Archives, 1925.

DeSoto, Clinton. *200 Meters & Down. The Story of Amateur Radio*. West Hartford, Connecticut: American Radio Relay League, Inc., 1936.

"Développement Du Réseau Radiotélégraphique Russe." *Journal Télégraphique* 4 (1912): 95.

"Diplomacy by Wireless." *Wireless World* (October 1918): 389–92.

Dobrjakov, Jurij. "The Case of the Peasant Zhidkovsky [Delo Krest'janina Zhidkovskogo]." *Radiofront* 13 (1935): 7–8.
Dotto, Simone. "Screening Radio or Broadcasting Cinema: How to Expose a Nameless Medium." In *A History of Cinema Without Names/3: New Research Paths and Methodological Glosses Udine*, 299–316. Milano – Udine: Mimesis Edizioni, 2018.
Douglas, Susan J. *Inventing American Broadcasting, 1899–1922*. Baltimore, Maryland: The Johns Hopkins University Press, 1989.
Douglas, Susan J. "Writing from the Archive: Creating Your Own." *Communication Review* 13, no. 1 (2010): 5–14. https://doi.org/10.1080/10714420903558613.
Douki, Caroline, and Philippe Minard. "Histoire Globale, Histoires Connectées: Un Changement d'échelle Historiographique." *Revue d'histoire Moderne et Contemporaine* 5, no. 54 (2007): 7–21.
Dunbar-Hester, Christina. ""Free the Spectrum!" Activist Encounters with Old and New Media Technology." *New Media and Society* 11, no. 1–2 (2009): 221–40. https://doi.org/10.1177/1461444808100160.
Dunlap, Orrin E. *Marconi, the Man and his Wireless*. New York: The Macmillan Company, 1937.
Duval, René. *Histoire de La Radio En France*. Paris: Alain Moreau, 1979.
Early morning [Rannee utro]. "News [Vesti]." 1916.
Eckersley, Peter Pendleton. "D. 40. N.8. Situation in Europe as Regards Wave-Lengths." In *Correspondance: Radiotélégraphie*. ITU Library & Archives, 1925.
Edwards, P.N., S. Jackson, M. Chalmers, G.C. Bowker, C.L. Borgman, D. Ribes, and S. Calvert. *Knowledge Infrastructures: Intellectual Frameworks and Research Challenges*. Ann Arbor, MI: University of Michigan School of Information, 2013.
Edwards, Paul N. *A Vast Machine: Computer Models, Climate Data, and the Politics of Global Warming*. Cambridge (MA), London: The MIT Press, 2010.
Edwards, Paul N. "Meteorology as Infrastructural Globalism." *Osiris* 21, no. 1 (2006): 229–50. https://doi.org/10.1086/507143.
Ellis, John. *Eye-Deep in Hell: Trench Warfare in World War I*. Baltimore: Johns Hopkins University Press, 1989.
Enrich, Juan Juliá. *Radio. Historia y Técnica*. Barcelona: Marcombo Boixareu Editores, 1993.
Ernst, Wolfgang. *Digital Memory and the Archive*. Edited by Jussi Parikka. Minneapolis, London: University of Minnesota Press, 2013. https://doi.org/10.5860/choice.50-5970.
Espagne, Michel. "La Notion de Transfert Culturel." *Revue Sciences/Lettres* 1, no. 1 (2013). https://doi.org/10.4000/rsl.219.
"Etats-Unis d'Amérique." *Journal Télégraphique* 4 (1912): 192.
"Farewell." *Electronics & Wireless World* 92 (1986): 60.
Fari, S., G. Balbi, and G. Richeri. "A Common Technical Culture of Telegraphy: The Telegraph Union and the Significance of Technological Standardization, 1865–1875." *HISTory of ELectro-Technology CONference, Third IEEE*. 2012.
Fari, Simone, Gabriele Balbi, and Giuseppe Richeri. *The Formative Years of the Telegraph Union*. Newcastle-Upon-Tyne: Cambridge Scholar Publishing, 2015.
Fauchille, P. "Le Domaine Aérien et Le Régime Juridique Des Aérostats." *Revue Générale de Droit International Public* 8 (1901): 414–85.
Felix, Edgar. "Department of Defence." *QST* February (1917): 20–21.
Fickers, Andreas. "Seeing the Familiar Strange: Some Reflections about Actants, Actors and Arenas of Transnational Media History." *Medien & Zeit. Kommunikation in Vergangenheit Und Gegenwart* 4 (2011): 16–24.

Fickers, Andreas. "Visibly Audible: The Radio Dial as Mediating Interface." In *The Oxford Handbook of Sound Studies*, 411–39. Oxford: Oxford University Press, 2012.

Fickers, Andreas, and Catherine Johnson. "Transnational Television History: A Comparative Approach." *Media History* 16, no. 1 (2010): 1–11. https://doi.org/10.1080/13688800903395411.

Fickers, Andreas, and Suzanne Lommers. "Eventing Europe: Broadcasting and the Mediated Performances of Europe." In *Materializing Europe: Transnational Infrastructures and the Project of Europe*, edited by Alexander Badenoch and Andreas Fickers. Basingstoke, Hampshire and New York: Palgrave Macmillan, 2010.

Finson, Jon W. *The Voices That Are Gone: Themes in 19th-Century American Popular Song*. New York, Oxford: Oxford University Press, 1994.

Fis, A. "Les Conférences Télégraphiques et Radiotélégraphiques de Madrid." *Bulletin d'informations, de Documentation et de Statistique. Ministère Des Postes, Télégraphes et Téléphones* 1 (1933): 13–21.

Fisher, Catherine. "World Citizens: Australian Women's Internationalist Broadcasts, 1930–1939." *Women's History Review* (2018): 1–19. https://doi.org/10.1080/09612025.2018.1506554.

Forlano, Laura. "WiFi Geographies: When Code Meets Place." *Information Society: An International Journal* 25, no. 5 (2009): 344–52. https://doi.org/10.1080/01972240903213076.

Friedewald, Michael. "Telefunken vs. Marconi, or the Race for Wireless Telegraphy at Sea, 1896–1914." In *SSRN Scholarly Paper*. Rochester, NY: Social Science Research Network, 2012.

"From the Editorial Board [Ot Redakcii]." *Herald of Military Radio Telegraphy and Electrical Engineering [Vestnik Voennoj Radiotelegrafii i Jelektrotehniki]* 1 (1917): 1–2.

Gallaix, Marcel G. de. "The Brussels Station. Wireless Telegraphy in War." *The Wireless World* (February 1915): 717–18. https://doi.org/10.2307/1323303.

Gamage, A.W. "Directory of Amateur Wireless Stations in the United Kingdom Licensed for Experimental Purposes by the Postmaster-General." 1913.

Gaudreault, A, and P. Marion. "Measuring the "'Double Birth'" Model against the Digital Age." *Early Popular Visual Culture* 11 (2013): 158–77.

General Post Office London. "D. 40. N. 22." *Correspondance: Radiotélégraphie*. ITU Library & Archives, 1925.

Gernshack, H. "The Radio League of America." *The Electrical Experimenter* December (1915): 381–83.

Gessler, Anne. ""Purifying the Upper Atmosphere": Women's Work in Early Radio, 1905–1913." *American Studies in Scandinavia* 46, no. 1 (2014): 87–102.

Giannini, Torquato Carlo. *La Radiotelegrafia Nell'economia e Nella Legislazione*. Roma: Agenzia Radiotelegrafica Italiana, 1920.

Gilfillan, Daniel. *Pieces of Sound. German Experimental Radio*. Minneapolis, London: University of Minnesota Press, 2009.

Given, Jock. "Born Global, Made Local: Multinational Enterprise and Australia's Early Wireless Industry." *Australian Economic History Review* 57, no. 2 (2017): 158–93.

Glushhenko, A.A. *Place and Role of Radio Communication in the Modernization of Russia (1900–1917) [Mesto i Rol' Radiosvjazi v Modernizacii Rossii (1900–1917)]*. Saint-Petersburg: A. S. Popov Naval Radioelectronics Institute, 2005.

Goodman, Jack, and Flora Carmichael. "Coronavirus: 5G and Microchip Conspiracies around the World." BBC News, 2020. https://www.bbc.com/news/53191523.

Gosar, Andrej. "Centennial National and Institutional Reports: Seismology and Physics of the Earth's Interior. 79.47. Slovenia." In *International Handbook of Earthquake and Engineering Seismology. Part B: Project of the Committee on Education, International Association of Seismology and Physics of the Earth's Interior*, edited by William Hung Kan Lee, Paul Jennings, Carl Kisslinger, and Hiroo Kanamori, 1431–32. Amsterdam, Boston et al.: Academic Press, 2003.

Greenstein, Shane. "Concentration in Internet Access and Entrepreneurial Truncation of Innovation." *Capitalism and Society* 7, no. 1 (2012): 1–31. https://doi.org/10.1515/1932-0213.1092.

Hagen, Wolfgang. *Das Radio: Zur Geschichte Und Theorie Des Hörfunks – Deutschland/USA*. Munich: Fink, 2005.

Hamilton, James F. "Excavating Concepts of Broadcasting: Developing a Method of Cultural Research Using Digitized Historical Periodicals." *Digital Journalism* 6, no. 9 (2018): 1136–49. https://doi.org/10.1080/21670811.2018.1481762.

Haring, Kristen. *Ham Radio's Technical Culture*. Cambridge (MA), London: The MIT Press, 2007.

Harley, J.B. "Deconstructing the Map." *Cartographica: The International Journal for Geographic Information and Geovisualization* 26, no. 2 (1989): 1–20.

Harley, J.B. "Maps, Knowledge, and Power." In *The Iconography of Landscape: Essays on the Symbolic Representation, Design and Use of Past Environments*, 277–313. Cambridge: Cambridge University Press, 1988.

Harlow, Alvin F. *Old Wires and New Waves*. New York: Arno Press and the New York Times, 1971.

Harvey, David. *The Condition of Postmodernity: An Enquiry into the Origins of Cultural Change*. Cambridge (MA), Oxford: Blackwell, 1989.

Haykin, Simon. "Cognitive Radio: Brain-Empowered Wireless Communications." *IEEE Journal on Selected Areas in Communications* 23, no. 2 (2005): 201–20. https://doi.org/10.1049/ir:20050504.

Headrick, Daniel R. *The Invisible Weapon: Telecommunications and International Politics, 1851–1945*. New York, Oxford: Oxford University Press, 1991.

Hendy, David. *Noise: A Human History of Sound and Listening*. London: Profile Books, 2013.

Hendy, David. "The Dreadful World of Edwardian Wireless." In *Moral Panics, Social Fears, and the Media: Historical Perspectives. Routledge Research in Cultural and Media Studies*. London and New York: Routledge, 2013.

Hendricks, John Allen, ed. *Radio's Second Century*. New Brunswick, Camden, and Newark, New Jersey, and London: Rutgers University Press, 2020.

Hepp, Andreas. "Transculturality As a Perspective: Researching Media Cultures Comparatively." *Forum Qualitative Sozialforschung/Forum: Qualitative Social Research* 10, no. 1 (2009).

Herkman, Juha. "Introduction: Intermediality as a Theory and Methodology." In *Intermediality and Media Change*, 10–29. Tampere: Tampere University Press, 2012.

Higgins, A. Pearce, ed. *The Hague Peace Conferences and Other International Conferences Concerning the Laws and Usages of War: Texts and Conventions with Commentaries*. London: Cambridge University Press, 1909.

Hilmes, Michele. *Network Nations: A Transnational History of British and American Broadcasting*. New York, Abingdon: Routledge, 2012.

Hilmes, Michele, and Jason Loviglio. *Radio Reader: Essays in the Cultural History of Radio*. London: Routledge, 2002.

Holl, U. "Radiophonie. Forschungen für ein kommendes Radio." *Historische Anthropologie. Kultur - Gesellschaft - Alltag*, 22(3) (2014): 426–435.

Hughes, Michael, and Katherine Bosworth. *Titanic Calling: Wireless Communications during the Great Disaster*. Edited by Michael Hughes and Katherine Bosworth. Oxford: The Bodleian Library, 2012.

Hugill, Peter J. *Global Communications since 1844: Geopolitics and Technology*. Baltimore, London: The Johns Hopkins University Press, 1999.

Huth, Arno. *La Radiodiffusion. Puissance Mondiale*. Paris: Gallimard, 1937.

Huurdeman, Anton A. *The Worldwide History of Telecommunications*. Hoboken: John Wiley & Sons, 2003.

ICRC. *Convention (V) Respecting the Rights and Duties of Neutral Powers and Persons in Case of War on Land. The Hague, 18 October 1907*. Hague: International Committee of the Red Cross, 1907.
"If We Are Closed Up." *QST* (April 1917): 26.
"Il Était Une Fois . . . Abel Gody." *T.S.F. Panorama* 14–15 (1991): 20–23.
"Il Telegrafo Senza Fili e La Meteorologia." *Rivista Delle Comunicazioni* (1911): 983–85.
"Illustrated London News: Greenwich Time from Paris," no. 3876 (2017): 3876.
"Inauguration Du Monument Commémoratif de La Fondation de l' Union Télégraphique." *Journal Télégraphique*, no. 12 (1922): 238–44.
"Interaction versus Consumption: Mass Media and Art from 1920 to Today." In *Timeshift: The World in Twenty-Five Years*. Ostfildern-Ruit: Hatje Cantz, 2004.
"International Amateur Radio Union." *The Wireless World* (July 1925): 46.
Iriye, Akira. "The Rise of Global and Transnational History." *Journal of Transnational American Studies* 5, no. 1 (2013).
Iriye, Akira, and Pierre-Yves Saunier. "Introduction: The Professor and the Madman." In *The Palgrave Dictionary of Transnational History. From the Mid-19th Century to the Present Day*, edited by Akira Iriye and Pierre-Yves Saunier, xvii–xx. New York: Springer, 2009.
ITU Archives. "Alfani. D. 19. N. 691." In *Registres de Correspondance: Radiotélégraphie*. 1913.
ITU Archives. *Carte Des Voies de Communcation Par TSF Entre Points Fixes*. 1927.
ITU Archives. "Carte Des Voies de Communication Par TSF Entre Points Fixes." Geneva: Bureau International de l'Union Télégraphique, 1927.
ITU Archives. "Carte Officielle Des Stations Radiotétégraphiques, 1° Édition, 1922–1923." 1923.
ITU Archives. "Circular No. 186." 1925.
ITU Archives. "Circular No. 752." 1923.
ITU Archives. "Convention Radiotélégraphique Internationale (1906)." In *Documents de La Conférence Radiotélégraphique Internationale (Berlin, 1906)*, 343–51. 1906.
ITU Archives. *Convention Radiotélégraphique Internationale (Londres, 1912)*. 1912.
ITU Archives. "D. 8. N. 69. Terré." In *Registres de Correspondance: Radiotélégraphie*. 1913.
ITU Archives. "D. 16. N. 1. Société Française Radioélectrique." In *Registres de Correspondance: Radiotélégraphie*. 1911.
ITU Archives. "D. 19. N. 15. Soc. Russe. de Télégr et Téléph. sans Fils." In *Registres de Correspondance: Radiotélégraphie*. ITU Library & Archives, 1917.
ITU Archives. "D. 19. N. 79. Soc. Russe. de Télégr et Téléph. sans Fils." In *Registres de Correspondance: Radiotélégraphie*. 1914.
ITU Archives. "D. 19. N. 87. Belar." In *Registres de Correspondance: Radiotélégraphie*. 1913.
ITU Archives. "D. 19. N. 95. Belar." In *Registres de Correspondance: Radiotélégraphie*. 1912.
ITU Archives. "D. 19. N. 97. Belar." In *Registres de Correspondance: Radiotélégraphie*. 1913.
ITU Archives. "D. 19. N. 114. Belar." In *Registres de Correspondance: Radiotélégraphie*. 1913.
ITU Archives. "D. 19. N. 409. Belar." In *Registres de Correspondance: Radiotélégraphie*. 1911.
ITU Archives. "D. 19. N. 691." *Registres de Correspondance: Radiotélégraphie*. 1913.
ITU Archives. "D. 21. N. 137. Belar." In *Registres de Correspondance: Radiotélégraphie*, 1911.
ITU Archives. "D. 21. N. 143. Belar." In *Registres de Correspondance: Radiotélégraphie*, 1911.
ITU Archives. "D. 21. N. 322. Terré." In *Registres de Correspondance: Radiotélégraphie*, 1913.
ITU Archives. "D. 21. N.43. Belar." In *Registres de Correspondance: Radiotélégraphie*. 1911.
ITU Archives. "D. 23. N. 1. Cie Franc Maritimes de t.s.f." In *Registres de Correspondance: Radiotélégraphie*. 1911.
ITU Archives. "D. 23. N. 6. Kümmerly and Frey." In *Registres de Correspondance: Radiotélégraphie*. 1911.
ITU Archives. "D. 23. N. 12. Falett Co." In *Registres de Correspondance: Radiotélégraphie*. 1911.

ITU Archives. “D.8. N. 33. Société Française Radioélectrique.” In *Registres de Correspondance: Radiotélégraphie*. 1919.
ITU Archives. “Documents de La Conférence Radiotélégraphique Internationale de Washington 1927. Tome I (Washington, 1927).” Bern: Bureau International de l’Union télégraphique, 1928.
ITU Archives. “Documents de La Conférence Radiotélégraphique Internationale de Washington 1927. Tome II (Washington, 1927).” Bern: Bureau International de l’Union télégraphique, 1928.
ITU Archives. *Documents de La Conférence Radiotélégraphique Internationale (Londres, 1912)*. Bern: Bureau International de l’Union télégraphique, 1913.
ITU Archives. *Documents de La Conférence Télégraphique Internationale de Paris, 1925, Tome II (Paris, 1925)*. Bern: Bureau International de l’Union télégraphique, 1926.
ITU Archives. *Documents of the Preliminary Conference on Wireless Telegraphy (Berlin, 1903)*. London: George Tucker, 1903.
ITU Archives. *International Radiotelegraph Convention (London, 1912)*. 1912.
ITU Archives. *International Radiotelegraph Convention of Washington, 1927 and General and Supplementary Regulations (Washington, 1927)*. London: His Majesty’s Stationery Office, 1928.
ITU Archives. *L’Union Télégraphique Internationale (1865 – 1915)*. Bern: Bureau international de l’Union télégraphique, 1915.
ITU Archives. “Proceedings at the Conference.” In *[Documents of the] Preliminary Conference on Wireless Telegraphy (Berlin, 1903). Translation of the Procès-Verbaux and Protocol Final by George R. Neilson*. London: George Tucker, 1903.
ITU Archives. “Rapport de Gestion de l’Union Télégraphique (Section Radiotélégraphique).” 1912.
ITU Archives. *Registres de Correspondance: Radiotélégraphie, 1912*. 1912.
ITU Archives. *Registres de Correspondance: Radiotélégraphie, 1913*. 1913.
ITU Archives. *Registres de Correspondance: Radiotélégraphie, 1914*. 1914.
ITU Archives. *Statistique Générale de La Radiotélégraphie Dressée d’après Des Documents Official. Année 1912*. Bern: Bureau International de l’Union télégraphique, 1914.
ITU Archives. *Statistique Générale de La Radiotélégraphie Dressée d’après Des Documents Official. Année 1914*. Bern: Bureau International de l’Union télégraphique, 1916.
ITU Archives. *What Is ITU?* Geneva: Bureau International de l’Union télégraphique, 1974.
ITU News. “Wi-Fi Takes the Sector by Storm.” *ITU News* 8, no. 8 (2003): 29–31.
Jenkins, Henry. *Confronting the Challenges of Participatory Culture: Media Education for the 21st Century*. Cambridge (MA), London: MIT Press, 2009.
Jenkins, Henry. *Convergence Culture: Where Old and New Media Collide*. New York and London: New York University Press, 2006.
Jenkins, Henry. *Fans, Bloggers, and Gamers: Exploring Participatory Culture. New York University Press*. New York, London: NYU Press, 2006.
Jenkins, Henry. *Textual Poachers: Television Fans and Participatory Culture*. New York, London: Routledge, 1992.
Jessop, George Robert. *The Bright Sparks of Wireless*. Radio Society of Great Britain, 1990.
John, Richard R. “When Techno-Diplomacy Failed: Walter S. Rogers, the Universal Electrical Communications Union, and the Limitations of the International Telegraph Union as a Global Actor in the 1920s.” *History of the International Telecommunication Union* (2020): 55–76. https://doi.org/10.1515/9783110669701-004.
Joly, Norman F. *The Dawn of Amateur Radio in the U.K. and Greece*. London: Project Gutenberg, 1990.
Jones, Reece. “Categories, Borders and Boundaries.” *Progress in Human Geography* 33, no. 2 (2009): 174–89. https://doi.org/10.1177/0309132508089828.

Kahn, David. *The Codebreakers: The Comprehensive History of Secret Communication*. New York: Scribner, 1996.

Keohane, Robert O., and Joseph S Jr. Nye. "Transnational Relations and World Politics: An Introduction." *International Organization* 25, no. 3 (1971): 329–49.

Kern, Stephen. *The Culture of Time and Space, 1880–1918: With a New Preface*. Cambridge (MA), London: Harvard University Press, 2003.

Kipling, Rudyard. "Wireless (1902)." In *Traffics and Discoveries*, 164–85. Cornwall: House of Stratus, 2008.

Kirpal, Alfred. "Ernst Oder Spiel? Basteln, Konstruieren Und Erfinden in Der Radioentwicklung." In *Homo Faber Ludens. Geschichten Zu Wechselbeziehungen von Technik Und Spiel*, 4: 227–54. Frankfurt am Main: Peter Lang, 2003.

Kocka, Jürgen. "Comparison and Beyond." *History and Theory* 42, no. 1 (2003): 39–44.

Körner, W.F. *Geschichte Des Amateurfunks: Seine Anfänge (1909–1963)*. Hamburg: FT-Verlag Rojahn and Kraft, 1963.

Krajewski, Markus. *World Projects: Global Information before World War I*. Minneapolis: University of Minnesota Press, 2014.

"La Station Radiotélégraphique Côtière de Tahiti (Océanie)." *Journal Télégraphique* 2 (1916): 47.

"La Télégraphie sans Fils et l'aérostation." *Journal Télégraphique* 7 (1909): 176.

"La Téléphonie sans Fil. Conférence Faite Par M. Le Lieutenant de Vaisseau Colin à La Société Internationale Des Electriciens." *Journal Télégraphique* 1 (1910): 3–9.

Lacey, Kate. "Ten Years of Radio Studies: The Very Idea." *The Radio Journal* 6, no. 1 (2008): 21–32. https://doi.org/10.1386/rajo.6.1.21/4.

"Ladies on Radiotelegraphic Work." *The Wireless World* (October 1917): 479.

"Language in the Making." *The Wireless World* March (1917): 927–28.

Larcher, Maurice. "La Guerre Turque Dans La Guerre Mondiale." *Les Archives de La Grande Guerre et de l'histoire Contemporaine*, no. 44 (1921): 1405–58.

Lasswell, Harold D. *Propaganda Technique in the World War*. New York: Peter Smith, 1938.

Latour, Bruno. *Pandora's Hope: Essays on the Reality of Science Studies*. Cambridge, MA: Harvard University Press, 1999.

Le Département des Postes de l'Empire d'Allemagne. *Documents de La Conférence Radiotélégraphique Internationale (Berlin, 1906)*. Berlin: Le Département des Postes de l'Empire d'Allemagne, 1906.

Lemstra, Wolter. "Leadership with 5G in Europe: Two Contrasting Images of the Future, with Policy and Regulatory Implications." *Telecommunications Policy* 42, no. 8 (2018): 587–611. https://doi.org/10.1016/j.telpol.2018.02.003.

Lemstra, Wolter, Vic Hayes, and John Groenewegen, eds. *The Innovation Journey of Wi-Fi: The Road to Global Success*. Cambridge: Cambridge University Press, 2011. https://doi.org/10.1017/CBO9780511666995.

Lendon. The Wireless Widow. *The Sketch* (1923, June 6): 467.

Le Petit journal. Supplément du dimanche. "Le Vingt-Cinq Ans de Tour Eiffel." April 19, 1914.

Le Petit Parisien. "Courrier Des Amateurs de T.S.F." May 11, 1924.

Le Queux, William. *The Spy Hunter*. London: C. Arthur Pearson, Ltd., 1916.

Leshhinskij, Vladimir Mikhailovich. *On the Nature of the Wireless Telegraph [O Sushhnosti Bezprovolochnago Telegrafa]*. Tver: B.I., 1918.

Lewis, J R. "Correspondence." *The Wireless World* (July 1916): 215.

Lindley, Dan. *Promoting Peace with Information: Transparency as a Tool of Security Regimes*. Princeton, Oxford: Princeton University Press, 2007.

Lippmann, Stephen. “Boys to Men: Age, Identity, and the Legitimation of Amateur Wireless in the United States, 1909–1927.” *Journal of Broadcasting and Electronic Media* 54, no. 4 (2010): 657–74. https://doi.org/10.1080/08838151.2010.519812.
“List of Subscribers to War News Sept: 1915.” University of Oxford, Bodleian Library, ms. Marconi 301, 1915.
Lo Biundo, Ester. *London Calling Italy: La Propaganda Di Radio Londra Nel 1943*. Milan: Unicopli, 2014.
Lochte, Bob. “U.S. Public Radio: What Is It – and for Whom?” In *More than a Music Box: Radio Cultures and Communities in a Multi-Media World*, 39–55, 1996.
Lockyer, William J.S. “International Time and Weather Radio-Telegraphic Signals.” In *Directory of Amateur Wireless Stations in the United Kingdom Licensed for Experimental Purposes by the Postmaster-General*, edited by A.W. Gamage, 3–6. London: A.W. Gamage, Ltd., 1913.
Lommers, Suzanne. *Europe – on Air: Interwar Projects for Radio Broadcasting*. Amsterdam: Amsterdam University Press, 2012.
Lovell, Stephen. *Russia in the Microphone Age. A History of Soviet Radio, 1919–1970*. Oxford: Oxford University Press, 2015.
“L’Union Radiotélégraphique.” *Journal Télégraphique* (juillet 1922): 124–26.
MacDonald, Simon. *Transnational History: A Review of Past and Present Scholarship*. London, 2015.
Mackenzie, Adrian. *Wirelessness. Radical Empiricism in Network Cultures*. Cambridge (MA), London: The MIT Press, 2010.
MacLennan, Anne F. “Learning to Listen: Developing the Canadian Radio Audience in the 1930s.” *Journal of Radio and Audio Media* 20, no. 2 (2013): 311–26. https://doi.org/10.1080/19376529.2013.825534.
Mansell, James G. *The Age of Noise in Britain: Hearing Modernity*. Urbana, Chicago, and Springfield: University of Illinois Press, 2017.
“Marconi “Amazons””. *The Wireless World* (March 1918): 841.
Marconi Archive. “Correspondence and Papers Concerning the Status of Various Firms and Industries during the War, 1910–18.” University of Oxford, Bodleian Library, MS. Marconi 3163, 1916.
Marconi Archive. “Documents Relating to the Imperial Wireless Scheme, 1911–15.” University of Oxford, Bodleian Library, MS. Marconi 219, 1915.
Marconi Archive. “Government -v- Private Control of Imperial Wireless Stations.” University of Oxford, Bodleian Library, ms. Marconi 301, 1916.
Marconi Archive. “Monthly Reports on News Services, 1916–21.” University of Oxford, Bodleian Library, Ms Marconi 302, 1917.
Marconi Archive. “MS Marconi 1798. HIS 150: Wireless Telephony – Early Tests to 1921.” 1921.
Marconi Archive. “Papers Concerning the National Status of the General Electric Company, 1914–16.” University of Oxford, Bodleian Library, MS. Marconi 3152, 20–25. 1916.
Marconi Archive. “Serial C. 3021. Press Bureau.” December 9, 1916.
Marconi Archive. “Wireless News Scheme. Europe [Hand-Drawn Map].” University of Oxford, Bodleian Library, ms. Marconi 301. 1920.
Marconi Calling Digital Archive. “SS Titanic (1).” Chelmsford, 1912. https://wayback.archive-it.org/org-467/20121016163838/http://www.marconicalling.com/museum/html/objects/photographs/large_image/large_image-type_d__t04582.html.
Marconi’s Wireless Telegraph Co. Ltd. (1920). Proposal for a network of wireless communications to serve the needs of the whole British Empire. In SC MSS 143/8 (Printed Papers). IET Archives.
Marks, Paul. “Kissing Device Lets You Send a Wireless Smooch.” *New Scientist* 215, no. 2873 (2012): 20.

McCrossen, Alexis. *Marking Modern Times. A History of Clocks, Watches, and Other Timekeepers in American Life*. Chicago and London: The University of Chicago Press, 2013.

McIlroy, J.A. “The History of the Greenwich Time Signal from 1924 to the Present Day.” *Engineering Science and Education Journal* 2, no. 6 (1993): 281. https://doi.org/10.1049/esej:19930079.

Mee, Arthur, ed. “The Unseen Telegraph Messenger Turning Electric Waves into Words.” In *The Children’ Encyclopaedia*, 3526–27. Lewisham, London: Educational Book Company, 1914.

Meyer, Victor. *L’ Union Internationale Des Télécommunications et Son Bureau*. Bern, 1946.

“Military Aviation.” *The Wireless World* 11 (1914): 516.

Miñana, Jesús Sánchez. *La Introducción a Las Radiocomunicaciones En España (1896 – 1914)*. Madrid: Fundación Rogelio Segovia para el Desarrollo de las Telecomunicaciones Ciudad Universitaria, 2004.

Mintz, Sidney. *Sweetness and Power: The Place of Sugar in Modern History*. Harmondsworth: Penguin Books, 1986.

“Monthly Miscellany.” *Marconigraph* 2, no. 20 (1912): 361–62.

“More Women for Wireless.” *The Wireless World* (February 1918): 768.

Morley, David. *Media, Modernity and Technology: The Geography of the New*. London and New York: Routledge, 2007.

Mosco, Vincent. *The Political Economy of Communication*. London: SAGE Publications, 2014.

Müller, Simone M. *Wiring the World: The Social and Cultural Creation of Global Telegraph Networks*. New York, Chichester: Columbia University Press, 2016.

Murphy, Kate. *Behind the Wireless: A History of Early Women at the BBC*. Basingstoke and New York: Palgrave Macmillan UK, 2016.

Natale, Simone. “A Cosmology of Invisible Fluids: Wireless, X-Rays, and Psychical Research around 1900.” *Canadian Journal of Communication* 36, no. 2 (2011): 263–75.

Natale, Simone. *Supernatural Entertainments: Victorian Spiritualism and the Rise of Modern Media Culture*. University Park, Pa.: Pennsylvania State University Press, 2016.

Natale, Simone. “Unveiling the Biographies of Media: On the Role of Narratives, Anecdotes, and Storytelling in the Construction of New Media’s Histories.” *Communication Theory* 26, no. 4 (2016): 431–49. https://doi.org/10.1111/comt.12099.

Natale, Simone, and Gabriele Balbi. “Media and the Imaginary in History: The Role of the Fantastic in Different Stages of Media Change.” *Media History* 20, no. 2 (3 April 2014): 203–18.

National Archives of the United States. *43.2.11 Records of the Preliminary International Conference on Electrical Communications*. 1920.

Negus, Keith. “Global Harmonies and Local Discords: Transnational Policies and Practices in the European Recording Industry.” *European Journal of Communication* 8, no. 3 (1993): 295–316. https://doi.org/10.1177/0267323193008003003.

Neiberg, Michael S. “Toward a Transnational History of World War I.” *Canadian Military History* 17, no. 3 (2012).

“New Zealand.” *The Wireless World* (February 1917): 69.

“Notes of the Month.” *Wireless World* (1914): 372–73.

“Notes of the Month.” *The Wireless World* (January 1915): 640–41.

“Nouvelles Installations de Télégraphie sans Fil.” *Journal Télégraphique* 12 (1916): 262.

Office International de Radiophonie. “D. 40. N. 5.” In *Correspondance: Radiotélégraphie*. ITU Library & Archives, 1925.

Ogle, Vanessa. *The Global Transformation of Time: 1870–1950*. Cambridge (MA), London: Harvard University Press, 2015.

“Our Article in the February Number . . .” *The Wireless World* (March 1916): 793.

Parks, Lisa. "Earth Observation and Signal Territories: Studying U.S. Broadcast Infrastructure through Historical Network Maps, Google Earth, and Fieldwork." *Canadian Journal of Communication* 38 (2013): 285–307.

Peck, Raymond W., and Robert Hood Bowers. *Love's Wireless Telephone [Song]*. Detroit, New York: Jerome H. Remick & Co, 1906.

Peresypkin, Ivan Terentievich. *Radio – a Powerful Means of Defense of the Country [Radio – Moguchee Sredstvo Oborony Strany]*. Moscow: Military Publishing House of the Ministry of Armed Forces of the USSR, 1948.

Peters, Benjamin. "And Lead Us Not into Thinking the New Is New : A Bibliographic Case for New Media History." *New Media and Society* 11 (1 & 2), no. 1/2 (2009): 13–30.

Pike, Robert M., and Dwayne R. Winseck. "The Politics of Global Media Reform, 1907–23." *Media, Culture & Society* 26, no. 5 (2002): 643–75. https://doi.org/10.1177/0163443704045505.

Pinch, Trevor J., and Wiebe E. Bijker. "The Social Construction of Facts and Artefacts: Or How the Sociology of Science and the Sociology of Technology Might Benefit Each Other." *Social Studies of Science* 14, no. 3 (1984): 419–24.

Piotrowski, Suzanne J, ed. *Transparency and Secrecy: A Reader Linking Literature and Contemporary Debate*. Lanham, Boulder, New York, Toronto, Plymouth: Lexington Books, 2010.

Plaiss, Adam. "Who Gets to Draw the Map? The Contentious Creation of the American Road/Map System, 1917–1926." *History and Technology* 28, no. 1 (2012): 3–24.

"Plan Showing Principal Stations of Proposed Press Association Wireless Scheme. Dwg No 16726." University of Oxford, Bodleian Library, ms. Marconi 301, 1920.

Potter, Simon J. *Broadcasting Empire: The BBC and the British World, 1922–1970*. Oxford: Oxford University Press, 2012.

Potter, Simon J. *Wireless Internationalism and Distant Listening: Britain, Propaganda, and the Invention of Global Radio, 1920–1939*. Oxford: Oxford University Press, 2020.

Potter, Simon J, and Jonathan Saha. "Global History, Imperial History and Connected Histories of Empire." *Journal of Colonialism and Colonial History* 16, no. 1 (2015): 1–14. https://doi.org/10.1353/cch.2015.0009.

Powell, Alison. "WiFi Publics: Producing Community and Technology." *Information Communication and Society* (2008). https://doi.org/10.1080/13691180802258746.

Printemps, Yvonne. "Concert à La Tour Eiffel Par TSF." *Agence Rol*, 1922.

"Progress in France." *The Wireless World* (May 1918): 91.

"Proposed American Radio Relay League Trunk Lines. " *QST* February (1916): 21.

"Questions and Answers." *The Wireless World* (September 1914): 409–10.

Raboy, Marc. *Marconi: The Man Who Networked the World*. Oxford: Oxford University Press, 2016.

Radio Station Marconi. "D. 40. N. 6." In *Correspondance: Radiotélégraphie*. ITU Library & Archives, 1925.

"Radiofication in the USSR [Radiofikazia SSSR]." *Radio Vsem* 7 (1927): 145.

"Radiotélégraphie Transpacifique." *Journal Télégraphique* 9 (1916): 216.

"Réseau Intercolonial Français de Télégraphie sans Fil." *Journal Télégraphique* 03 (1912): 72.

"Results of Russian Experiment." *The Wireless World* (February 1915): 686.

"Revue Télégraphique de 1911." *Journal Télégraphique* 1 (1912): 1–4.

"Revue Télégraphique de 1914." *Journal Télégraphique* 1 (1915): 1–4.

"Revue Télégraphique de 1915." *Journal Télégraphique* 1 (1916): 1–3.

"Revue Télégraphique de 1918." *Journal Télégraphique* 1 (1919): 1–3.

"Revue Télégraphique de 1922." *Journal Télégraphique* 1 (1923): 1–3.

"Revue Télégraphique de 1923." *Journal Télégraphique* 1 (1924): 1–4.

Ribeiro, Nelson. "Censorship and Scarcity: Controlling New and Old Media in Portugal, 1936–1945." *Media History* 21, no. 1 (2015): 74–88. https://doi.org/10.1080/13688804.2014.950951.
Richards, Elaine. *Radio Society of Great Britain Centenary. 100 Years Working for Amateur Radio*. Bedford: Radio Society of Great Britain, 2013.
Rikitianskaia, Maria. "How Children Learned to Listen: The Formation of Radio Clubs in the Soviet Union [Kak Detej Uchili Slushat'(Sja): Stanovlenie Radiokruzhkov v Sovetskom Sojuze] (in Russian)." *Logos* 27, no. 5 (2017): 141–62.
Rikitianskaia, Maria. "Listening to "Concert of Europe": Pioneering Radio Amateurs during World War I." Edited by Chris Hart. *World War I: Media, Entertainments & Popular Culture*, 2018, 122–42.
Rikitianskaia, Maria. "The International Radiotelegraph Union Facing World War I, 1912–1927." In *History of the International Telecommunication Union (ITU). Transnational Techno-Diplomacy from the Telegraph to the Internet*, edited by Gabriele Balbi and Andreas Fickers, 191–214. Berlin: De Gruyter, 2020.
Rikitianskaia, Maria. '"The Real Ethernet": The Transnational History of Global Wi-Fi Connectivity'. *New Media & Society* (2022): 1–20. https://doi.org/10.1177/14614448221103533.
Rikitianskaia, Maria, and Gabriele Balbi. "Radio Studies beyond Broadcasting: Towards an Intermedia Radio History." *Radio Journal: International Studies in Broadcast & Audio Media* 18, no. 2 (2020).
Rikitianskaia, Maria, Gabriele Balbi, and Katharina Lobinger. "The Mediatization of the Air. Wireless Telegraphy and the Origins of a Transnational Space of Communication, 1900–1910s." *Journal of Communication* 68, no. 4 (2018): *758–79.* https://doi.org//10.1093/joc/jqy030.
Rikitianskaia, Maria, Paolo Bory. "Images of global communication: revealing maps of telecommunications infrastructure [Obrazy global'noj kommunikacii: raskryvaem karty telekommunikacionnyh infrastruktur]" (in Russian). *NZ*, 130, no. 2 (2020): 40–54.
Ring, Justin. *Kiss Me by Wireless*. Ring-Hager, 1922.
Rollet, Laurent, and Philippe Nabonnand. "Why Aerodynamics Failed to Take off in Nancy: An Unexpected Casualty of World War I." In *A War of Guns and Mathematics*, edited by David Aubin and Catherine Goldstein, 351–69. Providence, R.I: American Mathematical Society, 2014.
Saint-Martin, Arnaud. "L'office et le télescope: une sociologie historique de l'astronomie française, 1900–1940." Université de Paris-Sorbonne, 2008.
Sarkar, Tapan K., Robert J Mailloux, Arthur A. Oliner, Magdalena Salazar-Palma, and Dipak L Sengupta. *History of Wireless*. New Jersey: Wiley-Interscience, 2006.
Satia, P. "War, Wireless, and Empire: Marconi and the British Warfare State, 1896–1903." *Technology and Culture* 51, no. 4 (2010): 829–53.
Scales, Rebecca P. *Radio and the Politics of Sound in Interwar France, 1921–1939*. Cambridge: Cambridge University Press, 2016.
Schiavone, Francesco. *Communities of Practice and Vintage Innovation: A Strategic Reaction to Technological Change*. Heidelberg, New York, Dordrecht, London: Springer Science & Business Media, 2014.
Schindler, Diethrich, and Jiri Toman. *The Laws of Armed Conflicts*. Leiden, Boston: Martinus Nihjoff Publisher, 2004.
Schipper, Frank, and Johan Schot. "Infrastructural Europeanism, or the Project of Building Europe on Infrastructures: An Introduction." *History and Technology* 27, no. 3 (2011): 245–64. https://doi.org/10.1080/07341512.2011.604166.
Schroeder, Peter B. "The Radio Amateur in International Legislation and Administration." *The American Journal of International Law* 48, no. 3 (1954): 421–33.
Schueler, Judith. *Materialising Identity: The Co-Construction of the Gotthard Railway and Swiss National Identity*. Amsterdam: Aksant, 2008.

"Scientific Time Signals." *The Wireless World* (August 1913): 299.

Seul, Stephanie, and Nelson Ribeiro. "Revisiting Transnational Broadcasting: The BBC's Foreign-Language Services during the Second World War." *Media History* 21, no. 4 (2015): 365–77.

Ribeiro, Nelson, and Seul, Stephanie (Eds.). *Revisiting transnational broadcasting: the BBC's foreign-language services during the Second World War*. London, New York: Routledge, 2017.

Shah, Rajiv C., and Christian Sandvig. "Software Defaults as de Facto Regulation: The Case of the Wireless Internet." *Information Communication and Society* 11, no. 1 (2008): 25–46. https://doi.org/10.1080/13691180701858836.

Shamshur, Vladimir Ivanovich. *The First Years of Soviet Radio Engineering and Radio Amateurs [Pervye Gody Sovetskoj Radiotehniki i Radioljubitel'stva]*. Moscow, Leningrad: Gosjenergoizdat, 1954.

Sherwood, Marika. "Strikes! African Seamen, Elder Dempster and the Government 1940–42." *Immigrants & Minorities: Historical Studies in Ethnicity, Migration and Diaspora* 13, no. 2–3 (1994): 130–45.

Siemens Brothers. "D. 4. N. 74." *Registres de Correspondance: Radiotélégraphie*. ITU Library & Archives, 1915.

Siemens Brothers. "D. 19. N. 799." In *Registres de Correspondance: Radiotélégraphie*. ITU Library & Archives, 1913.

Siemens Brothers. "D.4. N. 165–167." *Registres de Correspondance: Radiotélégraphie*. ITU Library & Archives, 1914.

Silva, Ana Paula. "From Point-to-Point to Mass Communication: The Radio in Portugal from 1898 to 1939." Lisbon: Institute of Social Sciences, University of Lisbon, 2010.

Silverstone, Roger, and Erik Hirsch, eds. *Consuming Technologies: Media and Information in Domestic Spaces*. London: Routledge, 1992.

Simeon, Michel. "Les Télécommunications et La Guerre de 1914 à 1918. Avènement de La T.S.F." ITU Library & Archives, 1989.

Sismondo, Sergio. "Science and Technology Studies and an Engaged Program." In *The Handbook of Science and Technology Studies*, edited by E. Hackett, O. Amsterdamska, M. Lynch, and Judy Wajcman, 13–32. Cambridge (MA), London, 2008.

Some Essex Industries. The Romance of Marconi's wireless wonders at Chelmsford. Essex Weekly News, Friday, August 23, 1935. Marconi Archive. Ms Marconi 200.

Spohrer, Jennifer. "Ruling the Airwaves: Radio Luxembourg and the Origins of European National Broadcasting, 1929–1950." Columbia University, 2008.

Starkey, Guy. *Local Radio, Going Global*. Basingstoke and New York: Palgrave Macmillan, 2011.

Starosielski, Nicole. *The Undersea Network*. Durham: Duke University Press, 2015.

Starosielski, Nicole, and Lisa Parks, eds. *Signal Traffic: Critical Studies of Media Infrastructures*. University of Illinois Press, 2015.

Strachan, Hew. "The First World War as a Global War." *First World War Studies* 1, no. 1 (2010): 3–14. https://doi.org/10.1080/19475021003621036.

Streeter, Thomas. *Selling the Air: A Critique of the Policy of Commercial Broadcasting in the United States*. Chicago, London: University of Chicago Press, 1996.

"Sunday, October 14th." *Wireless World* 10 (1928): 507.

"Teaching Device for Radiotelegraphists [Pribor Dlja Obuchenija Radiotelegrafistov]." *Herald of Wireless Telegraphy [Vestnik Telegrafii Bez Provodov]* (1913): 101.

Tekniska Museet. "Telefontornet i Rimfrost, 1890-Talet. Stockholms Allmänna Telefon." 1890. https://digitaltmuseum.se/021016458638/telefontornet-i-rimfrost-1890-talet-stockholms-allmanna-telefon-ab.

"Telephonic Time Service in Switzerland." *The Wireless Age* March, 5, no. 6 (1918): 488.

"The Admiralty Have Just Ordered . . ." *The Engineer* 6 (1901): 666.
The day book. (Chicago, Ill.). "Story in Picture of How Wireless Waked the Midnight Sea." April 17, 1912.
"The Day That Is Coming." *My Children's Magazine*, October 1914, 135–43.
"The Early Days of Wireless Telegraphy. How a Chess Match Was Played." *The Wireless World* 6 (1913): 202–3.
"The Effects of the War." *The Wireless World* (1914): 355.
"The Enlightener. It Is Necessary to Help School [Prosveshhenec. Nado Pomoch' Shkole]." *Bjulleten' Radio* 1 (1925): 20–21.
"The Experiences of an Amateur." *Wireless World* 1, no. 8 (1913): 507.
The Institute of International Law. "Project Drafted by the Institute of International Law for the Regulation of Aerostats and Wireless Telegraphy (1906)." *The American Journal of International Law* 7, no. 3, Supplement: Official documents (1913): 147–49.
"The Ladies Are Coming." *QST* August (1917): 19.
"The Library Table." *The Wireless World* 6 (1919): 585–87.
"The Quenching of Sayville." *The Wireless World* (November 1915): 515–17.
"The Radio Ladies." *QST* (March, 1921): 30.
"The Radio Ladies (Concluded)." *QST* (March, 1921): 64.
"The Report of the Civil Aerial Transport Committee. Appendix A. International Convention in Regard to Aerial Navigation Drafted by the Conference Held at Paris in 1910." *Flight* (1918): 1465–79.
"The Signal service. Wireless Telegraphy in the war." *The Wireless World* (February 1915): 718–721.
The Times. "The Marconi Agreement." August 8, 1912.
"The Wireless Press Limited Established to Carry on the Business of Publishers . . ." University of Oxford, Bodleian Library, ms. Marconi 301, 1916.
"The Wireless Spy." *The Wireless World* (January 1915): 646.
"The Young Man's Eyes Fell on His Own Transmitter Key. It Was Clicking up and Down." *The Wireless World* 1 (1915): 639.
Thibault, Ghislain. "Bolts and Waves: Representing Radio Signals." *Early Popular Visual Culture* 16, no. 1 (24 May 2018): 39–56. https://doi.org/10.1080/17460654.2018.1472621.
Thibault, Ghislain. "Wireless Pasts and Wired Futures." In *Theories of the Mobile Internet: Materialities and Imaginaries*, edited by Andrew Herman, Jan Hadlaw, and Tom Swiss, 87–104. New York, Oxon: Routledge, 2015.
Thompson, J.B. *The Media and Modernity: A Social Theory of the Media*. Stanford, CA: Polity Press, 1995.
Thrower, Keith R. "Technical Factors Affecting CW Radio Communication in WW1." Innovating in Combat. Telecommunications and intellectual property in the First World War. Reading, 2014.
Thussu, Daya Kishan. "Mapping Global Media Flow and Contra-Flow." In *Media on the Move: Global Flow and Contra-Flow*, edited by Daya Kishan Thussu. London and New York: Routledge, 2007.
"Time Signals." *The Marconigraph* (1912): 315.
""Titanic" Disaster. Hearings before a Subcommittee of the Committee on Commerce, 62 Congress, 2 Session, Pursuant to S. Res. 283, Directing the Committee on Commerce to Investigate the Causes Leading to the Wreck of the White Star Liner "Titanic"." United States Senate, 1912.
Toffler, Alvin. *The Third Wave*. London: Collins, 1980.
Topik, Steven C., and Allen Wells. *Global Markets Transformed, 1870–1945*. Boston, MA: Harvard University Press, 2012.
Torrens, Paul M. "Wi-Fi Geographies." *Annals of the American Association of Geographers* 98, no. 1 (2008): 59–84.
"Training Aviators." *The Wireless World* (July 1917): 234–35.
Tucker, Spencer. *The Great War, 1914–18*. Bloomington: Indiana University Press, 1998.

Tworek, Heidi J.S. “A Union of Nations or Administrations? Voting Rights, Representation, and Sovereignty at the International Telecommunication Union in the 1930s.” In *History of the International Telecommunication Union*, edited by Gabriele Balbi and Andreas Fickers, 243–64. 2019.

Tworek, Heidi J.S. “How Not to Build a World Wireless Network: German–British Rivalry and Visions of Global Communications in the Early Twentieth Century.” *History and Technology* 32, no. 2 (2016): 178–200. https://doi.org/10.1080/07341512.2016.1217599.

Ulman-Mauriat, Caroline. *Naissance d'un Média. Histoire Politique de La Radio En France (1921 - 1931)*. Paris, Montreal: L'Harmattan, 1931.

Union International de Radiophonie. “Stations de Radiodiffusion Emettant Sur Des Longueurs d'onde de 200 a 600 m.” In *Correspondance: Radiotélégraphie. 40*. ITU Archives, 1926.

Valderhaug, Gudmund. “Recordkeeping in Local Government in Norway 1950 - 2000.” *Archival Science* 3 (2003): 205–12.

Van der Vleuten, Erik. “Toward a Transnational History of Technology: Meanings, Promises, Pitfalls.” *Technology and Culture* 49, no. 4 (2008): 974–94.

VanCour, Shawn Gary. “The Sounds of “Radio”: Aesthetic Formations of 1920s American Broadcasting (PhD. Thesis).” University of Wisconsin-Madison, 2008.

Vleuten, Erik van der, and Arne Kaijser. “Networking Europe.” *History and Technology* 21, no. 1 (2005): 21–48. https://doi.org/10.1080/07341510500037495.

Wajcman, Judy. *Pressed for Time. The Acceleration of Life in Digital Capitalism*. Chicago and London: The University of Chicago Press, 2015.

Wakeford, Nina. “Networking Women and Grrrls with Information/Communication Technologies: Surfing Tales of the World Wide Web.” In *Processed Lives: Gender and Technology in Everyday Life*, edited by J. Terry and M. Calvert, 51–66. London and New York: Routledge, 1997.

Walker, Jesse. *Rebels on the Air: An Alternative History of Radio in America*. New York and London: New York University Press, 2001.

Wani, Kentaro. *Neutrality in International Law: From the Sixteenth Century to 1945*. London and New York: Routledge, 2017.

“War?” *QST* (March 1917): 29.

“War!” *QST* (May 1917): 1–3.

“War Notes.” *The Wireless World* (November 1914): 505.

Warner, Kenneth B. “Editorial. With the International Radiotelegraph Conference Safely behind Us.” *QST*, 1928.

Warner, Kenneth B. “The Amateur and the International Radiotelegraph Conference.” *QST* (January 1928): 15–22.

Washington gets Paris time, January The Wireless Age § (1914).

Wasko, Janet. “The Study of the Political Economy of the Media in the Twenty-First Century.” *International Journal of Media & Cultural Politics* 10, no. 3 (2014): 259–71. https://doi.org/10.1386/macp.10.3.259.

Wasson, Haidee. “Protocols of Portability.” *Film History* 25, no. 1–2 (2013): 236–47. https://doi.org/10.2979/reseafrilite.47.2.09.

Waterson, Jim, and Alex Hern. “At Least 20 UK Phone Masts Vandalised over False 5G Coronavirus Claims.” *The Guardian*, April 6, 2020. https://www.theguardian.com/technology/2020/apr/06/at-least-20-uk-phone-masts-vandalised-over-false-5g-coronavirus-claims.

Wehler, Hans-Ulrich. “Transnationale Geschichte — Der Neue Königsweg Historischer Forschung?” In *Notizen Zur Deutschen Geschichte*, 61–73. München: Verlag C.H. Beck, 2007.

Werbach, Kevin. *Radio Revolution: The Coming Age of Unlicensed Wireless*. Washington: New America Foundation Public Knowledge, 2003.
Werner, Michael, and Bénédicte Zimmermann. "Beyond Comparison: Histoire Croisée and the Challenge of Reflexivity." *History and Theory* 45, no. 1 (2006): 30–50.
Werner, Michael, and Bénédicte Zimmermann. "Penser l'histoire Croisée: Entre Empirie et Réflexivité." *Annales. Histoire, Sciences Sociales* 58, no. 1 (2003): 7–36.
"When a Woman Walks through a Wireless Telephone Wave." *The Wireless Age* December (1915): 210.
Whitehead, Gregory. "Out of the Dark. Notes on the Nobodies of Radio Art." In *Wireless Imagination: Sound, Radio and the Avant-Garde*, edited by Douglas Kahn and Gregory Whitehead, 253–65. Cambridge (MA), London: The MIT Press, 1992.
Wijfjes, Huub. "Spellbinding and Crooning: Sound Amplification, Radio, and Political Rhetoric in International Comparative Perspective, 1900–1945." *Technology and Culture* 55, no. 1 (2014): 148–85. https://doi.org/10.1353/tech.2014.0013.
Williams, Raymond. *Television: Technology and Cultural Form*. New York: Schocken Books, 1975.
Wilson, Woodrow, Gerhard Peters, and John T. Woolley. "Executive Order 2011—To Enforce Neutrality of Wireless Stations." *The American Presidency Project*, August 5, 1914. http://www.presidency.ucsb.edu/ws/?pid=75364.
Winseck, Dwayne R., and Robert M. Pike. *Communication and Empire: Media, Markets, and Globalization, 1860–1930*. Durham: Duke University Press, 2007.
Winseck, Dwayne R., and Robert M. Pike. "Communication and Empire. Media Markets, Power and Globalization, 1860–1910."*Global Media and Communication* 4, no. 1 (2008): 7–36.
"Wireless Amateurs in England Are . . ." The Wireless World, no. 3–33 (December 1915): 595.
"Wireless, Adj. and n." Oxford English Dictionary. Oxford University Press, 2020. https://www-oed-com.proxy.sbu.usi.ch/view/Entry/229458.
"Wireless and Women." *Truth*, April 14, 1912, 11.
"Wireless for Women." *The Wireless World* (March 1917): 928–29.
"Wireless in Air Navigation." *Wireless World* 2 (1919): 623–27.
"Wireless in the Courts." *The Wireless World* 3, no. 32 (1915): 523.
"Wireless in the War." *The Wireless World* (1914): 375–82.
"Wireless Police." *The Wireless World* (November 1914): 519–20.
"Wireless Signals for the Home." *The Wireless World* (March 1916): 794–97.
"Wireless Telegraphy for Women." *The Wireless World* (February 1916): 725–26.
"Wireless Weather Report." *The Wireless World* (November 1914): 522.
"Woman and Wireless." *Evening News*, December 4, 1912, 8.
"Women and War-Time Wireless." *The Wireless World* (November 1917): 545.
"World Wide Wireless." *The Wireless Age* 6, no. 9 (1919): 7–10.
"Worries of Wartime." *The Wireless World* 3 (1915): 800.
Wenzlhuemer, Roland. *Connecting the Nineteenth-Century World: The Telegraph and Globalization*. Cambridge: Cambridge University Press, 2012.
Wythoff, Grant. "Pocket Wireless and the Shape of Media to Come, 1899–1922." *Grey Room* 14, no. 51 (2013): 40–63. https://doi.org/10.1162/GREY_a_00106.
Zeller, R. Die EBU – Union Européenne de Radio-Télévision (UER) – European Broadcasting Union (EBU). Internationale Rundfunkkooperation Im Wandel. Baden-Baden: Nomos-Verlag, 1999.

Appendices

Appendix A. Total number of radio stations, 1912–1927

The current table represents the total number of radio stations from 1912 to 1927 (including ship and coastal stations) by country in decreasing order beginning with the information from 1912. The grey table cells indicate missing data. The data were extracted from a collection of the statistics from 1912 to 1927 published by the ITU Bureau. The data in the original statistics are often incomplete and have been supplemented with numerous additional notes and details; for any of these supplements, please refer to the primary source (ITU Archives, 1914b, 1915b, 1916, 1917b, 1918, 1919b, 1920, 1921, 1922, 1924, 1925b, 1925c, 1927c, 1928d, 1930). All the names of the countries have been entered as is, even in the case of changes in the country throughout this period (e.g. Russia and the Union of Soviet Socialist Republics are listed separately).

Open Access. © 2024 the author(s), published by De Gruyter. (cc) BY-NC-ND This work is licensed under the Creative Commons Attribution-NonCommercial-NoDerivatives 4.0 International License.
https://doi.org/10.1515/9783111202327-008

State	1912	1913	1914	1915	1916	1917	1918	1919	1920	1921	1922	1923	1924	1925	1926	1927
United States America	873	978	1138	1023	1166					973	289	837	74	125		
Great Britain	845	1275							3960	3928	3870	3887	3997	4030	4024	4049
Germany and Protectorates	399	546						83	92	220	557	724	766	770	814	920
France (Metropolitan)	246	301	369					14	1039	1100	1155	1171	1404	1366	1504	1521
Italy (Metropolitan)	142	177						311	243	499	747	834	813	916	660	888
Russia with Possessions and Protectorates	104	131	66	65	72					11						
Canada	97	122						245	278	302	306	311	313	329	356	368
Brazil	89			86	99	111	145						125		187	182
Netherlands	80	103	123	131	154	156	136	240	389	459	461	453	452	451	457	457
Austria	67	32	34													
Spain	66	46	79	80	60	109	400	175	200	236	498	513	400	339	438	459
Sweden	48	54	66	80	87	98	102	127	218	250	276	289	294	310	327	350
Japan	34	45	58	80	110	153	242	288	361	428	597	690	730	793	934	1097
Denmark	32	36	50	65	74	74	101	129	182	280	294	320	341	358	363	380
Greece	29	34	47	44							200	283	283			
Australia	26	44							123	165	184	193	183	158	152	
Belgium	19	20						54		125	129	142	128	133	144	143

New Zealand	19	27						41	42	27	6	30	43	64	86	8
British India	10	11	1					31	37	51	56	13	59	50	52	5
Italian Somalia	8	9	5									8	8	10	10	
Netherlands East-Indies	6	6	6	8	6	6	6	5	54	87	49	59	67	90	93	104
Romania	6	6	6					7	10		15	15	15	14	14	13
French West Africa	5	6		6				5		5	7	7	7	7		7
Portugal	5	24		33	40	46		50	58	58	81	119	93			
French Indochina	4	5	4	5	5	5	11	14	16	16	9	9	10	11	11	12
Morocco	4		8						4	3	4	5	4			
Madagascar	3	3	5	4	4	4		4		4	4	4	6	6	6	6
Curacao	3	3	3	3	3	3	2	2	1	3			9	13	22	
South Africa (Union)	2	3						2	2	3	12	12	14	14	18	19
British Somaliland	2	3										2	2	2	2	2
Trinidad and Tobago	2	2											4	2	2	
Egypt	1	1	1					2	2	6	6	7	8	2	2	12
Tunisia	1	2	2	2				2	2	3	3	3	3	3	3	3
Ceylon	1	1										2	1	2	2	2
Belgian Congo	1		1	1	1	1	1	1	1	1	1	1	1	1	1	1
Jamaica	1	2														

(continued)

(continued)

State	1912	1913	1914	1915	1916	1917	1918	1919	1920	1921	1922	1923	1924	1925	1926	1927
French Equatorial Africa	1	1	1	1	1	1	1		1	1		1	1	1	1	
Monaco	1	1	1													
British Guiana	1	1														
Norway		56	71	93	92	122	126	143	158	506		557	619	662	677	719
Union of Soviet Socialist Republics												73	124	132	143	159
Argentina (Republic)											108	109	116			138
Finland									13	23	23	22	28	42	46	63
Hong Kong													2	49	58	56
Iceland								3		11	12	21	32	46	51	50
Mexico		8				12	13	14		14	15	15	14	29	31	37
Estonia												27	28	29	32	28
Ireland (Free State of)													37	30	26	25
Danzig (Free City)										9	14	18	13	15	19	15
Mozambique											10	9	9		13	15
Cape Verde Islands												5	6		6	9
British West Indies															8	8

Ecuador															7	7
Angola																7
New Caledonia				3						1		4	4	3	3	7
Portuguese Guinea												2			3	3
Macao										1			3	3	3	3
French Establishments in Oceania					1			1	2	1	1	1	1	2	3	3
Martinique					3			1	2	2	2	2	2	2	2	2
S. Pierre and Miquelon									1			1	2	2	2	2
Reúnion												1	1	1	1	2
Bahamas Islands		1							70			1	1		4	1
Kenya										2		2	2	1	1	1
Tonga Islands													1	2	1	1
Guadeloupe								1	1	1	1	1	1	1	1	1
Czechoslovakia									1	1	1	1	1	1	1	1
French Somaliland									1	1	1	1	1	1	1	1
Gold Coast		1							1	1		1		1	1	1
British Honduras												1	1	1	1	1
French Guiana												1	1	1	1	1
New Hebrides										1	1	1			1	1

(continued)

(continued)

State	1912	1913	1914	1915	1916	1917	1918	1919	1920	1921	1922	1923	1924	1925	1926	1927
Cameroon													1	1	1	1
Lithuania														1	1	1
Territories of the Levant under French Mandate														1	1	1
S. Thomas															1	1
China								50	50	50	26	59				
Chile			37	35	29	7	9	34	75	47	72		87			
Yugoslavia															43	
Philippines												42	39	39		
Cuba													33			
Peru															26	
Latvia										3	14	19	21	25	36	
Italy (Colonies and Possessions)										14						
Uruguay		9	9													
Poland												7				
Hungary		1									7	7	7	7	7	

Straits Settlements												6	7	7	4	
Sarawak											5	9		2		
Fiji Islands		3										6	5	6		
Suriname														5	5	
British North Borneo												4	4	4	4	
Eritrea												4		4	4	
New Guinea									4	4						
Alaska					4											
Cyrenaica			3					3				4	4	4	4	
Portuguese East Africa and Asian Possessions				1		3				7						
Siam											3				3	
Gilbert and Ellice (Islands)													2	2	3	
Mauritius Island													2	2	3	
Zanzibar		2												2	2	
Dominican Republic												2				
Hawaii												2				
Ivory Coast													2			
Solomon Islands									2	1			1			

(continued)

(continued)

State	1912	1913	1914	1915	1916	1917	1918	1919	1920	1921	1922	1923	1924	1925	1926	1927
Tripolitania								1				1	1	1	2	
Gambia (Bathurst)													1	1	1	
Bulgaria			1							1						
Dahomey													1	1		
Greater Lebanon												1	1			
Marshall Islands									1	1						
New Brittany									1	1						
Admiralty (Islands of)										1						
Cocos-Keeling (Islands)		1														
Falkland Islands		1														
New Ireland										1						
Sierra Leone		1														
Tanganyika (Territory of)															1	
Togo (Territory under French Mandate)															1	

Index

Open Access. © 2024 the author(s), published by De Gruyter. BY-NC-ND This work is licensed under the Creative Commons Attribution-NonCommercial-NoDerivatives 4.0 International License.
https://doi.org/10.1515/9783111202327-009